Estimating Manufacturing Costs

Estimating Manufacturing Costs

A PRACTICAL GUIDE FOR MANAGERS AND ESTIMATORS

Lawrence M. Matthews

Certified Management Consultant

McGraw-Hill Book Company

New York · St. Louis · San Francisco · Auckland · Bogotá
Hamburg · Johannesburg · London · Madrid · Mexico
Montreal · New Delhi · Panama · Paris · São Paulo
Singapore · Sydney · Tokyo · Toronto

HD
47
·M36
1983

089915

Library of Congress Cataloging in Publication Data

Matthews, Lawrence M.
  Estimating manufacturing costs.
  Bibliography: p.
  Includes index.
  1. Costs, Industrial—Estimates. I. Title.
HD47.M36   658.1'552   82-15360
ISBN 0-07-040951-X   AACR2

1234567890DODO898765432

ISBN 0-07-040951-X

ISBN 0-07-040951-X

The editors for this book were William Sabin and Elizabeth P. Richardson,
the designer was Jules Perlmutter (Off-Broadway Graphics), and the production
supervisor was Paul Malchow. It was set in Melior by Bi-Comp, Incorporated.

Printed and bound by R. R. Donnelly and Sons Company.

To Ann, Jim, Barbara, Mark, and Larry

Contents

x □ **Contents**

Preface

Today management has greater and greater interest in improving the company's estimating of manufacturing costs. This statement is based on 35 years of management consulting experience and 137 management seminars given over the last 10 years before 3063 people from 2376 different companies.

Some managements are still guilty of such illogical statements as these: "We know our product costing isn't right, but look at that bottom line." "It's really not all that important that we do a good estimating job. When it's all said and done, it's competition that determines the prices we can charge." But such illogical attitudes are encountered less and less frequently. The reasons will be discussed in this book. Increasing interest in doing a better estimating job is a fact of modern management life.

Estimating is a difficult job, and a very individualistic one. Too little training and help are available in this management area. Our business schools deal with it lightly, if at all. Because of its confidentiality and possible governmental implications, trade associations are of little help. Occasional magazine articles are too rudimentary to be of much use to an estimator dealing with a specific product situation. Therefore there is a need for more exposition and thought on estimating manufacturing costs. There is also a need for a clearer understanding of the responsibilities of Estimating and its interrelationships with other management areas, i.e., the management aspects of estimating. These are all reasons for this book. It is written not for the estimator alone but also for the senior managers who either oversee Estimating's work or make decisions on the data and costs developed by Estimating.

Estimating is clearly a very individualistic matter, varying between industries and even between products within an industry. Nevertheless certain basic understandings, requirements, and techniques are almost universal, whether we are making biscuits, fountain pens, pumps, or airframes. They warrant our perception, acceptance, and use if we are to do a better estimating job. For example, there are certain basic steps involved in any estimate that must be recognized and followed.

The importance of a good estimate goes almost without saying. And yet it must be said. The company's ultimate success, even survival, depends on good estimates and good product costing. In the face of that plain fact, you will encounter managements with obvious and admittedly poor costing that will enter into major contracts and sales agreements without a good grasp of their product's cost. You think of managers as human beings with warm red blood in their veins. Managers who do this must have ice water in their veins! But, you see it. Even worse, you will find managements that actively resist a good costing effort because of the cost facts it will reveal on one or more favorite products. This may sound like an extreme statement, but it has been backed up by many practicing estimators at my estimating seminars.

This book and I owe a great deal to these seminar participants. No one, even after 35 years of experience, can have seen it all. But dealing with this estimating problem not only with clients but also with so many practitioners at so many seminars has developed a grasp of the subject that reflects their experience as well as my own. I am greatly indebted to the men and women who reacted, discussed, and revealed their own problems and solutions in this important management area. This book reflects their experience as well as mine, and I thank them for sharing it.

It is one of the goals of every author to have the reader say "That's just the way it happens. I've lived through it myself." I hope that will be my readers' reaction.

Lawrence M. Matthews

Estimating Manufacturing Costs

Basic Understanding in Estimating

Before we can look at any specific estimating situations or techniques of estimating we must define what an estimate is and relate it to product costing and pricing. We shall consider some common estimating difficulties, not to be negative but to help define the problems many of us face and the basic understanding management needs for good estimating. Then we can delineate the factors that are making estimating more difficult even while its true importance receives more recognition.

Definition of an Estimate

An estimate is an independent, realistic prediction of what it will cost to make a given product. In a service company, it is an independent, realistic prediction of what it will cost to render a given service. Let us examine this definition in detail.

The estimate must be independently arrived at. This means that it must be unbiased, free from whim, and unaffected by emotion or preconceived preferences. This characteristic of the estimate has some subtle and important implications. For example, Marketing (or Sales) in its wish for greater volume and its effort to push a given product can attempt to impose its judgment and influence to color an estimate. This matter will be dealt with at greater length when we consider Estimating's place in the company organization.

The estimate must be realistic. It must reflect the estimator's considered judgment of what can actually be expected. The estimate must therefore be in the often narrow zone between optimism and pessimism, insofar as the estimator can identify that zone. This characteristic of the estimate, like its independence, can have important results. For example, Production may exert pressures on Estimating that result in an unrealistically optimistic prediction of costs.

And finally, an estimate is a prediction. It is a view of the future. It may be, and often necessarily is, based largely on past performances and costs, but it is still a prediction and it still must reflect the considered judgment of the product-cost estimator, independently and realistically arrived at.

Estimating vis-à-vis Product Costing

Some people make a distinction between product-cost estimating and product costing that sounds reasonable and clear at first. For example, in many companies Estimating develops an estimate of the costs to make a given new product or new model of an existing product. In turn, Cost Accounting collects actual costs and calculates the current cost of making existing products. Thus, these people say that Estimating deals with new product or model costing and Cost Accounting deals with present product costing; i.e., they are separate and distinct matters.

I believe that this distinction is wrong and dangerous. It is wrong because it implies that Estimating has no concern for, and need have no involvement with, costing existing products, being concerned only with predicting costs of new products or modifications of existing products. That seems too narrow a definition of Estimating's true function. It is also a dangerous distinction because applying it means that costing existing products is done only on the basis of past actual costs and performance, which is too narrow an approach to take.

It is important to know past actual product costs. For example, we need those data for estimating follow-up. Finance (or Accounting) needs it for inventory valuations. But we also have to estimate what those existing products will cost us in the future, as we continue to make them or the next time we make them. This is the function of Estimating.

In my view, estimating and product costing are one and the same thing. Estimating should be considered responsible for estimating the cost of both new and existing products. In this book estimating and product costing are synonymous.

Estimating versus Pricing

Even experienced estimators make statements that indicate a confusion between estimating and pricing. They are two completely different matters.

Obviously, in the long run, prices must reflect costs. If they do not, there will be a new management—or no company. But in real life the price that can be charged may not include the full costs and the desired profit markup. This harsh situation has nothing to do with estimating. *Estimating* is the realistic prediction of what it will cost to make the product, and it is done by the Estimating department. *Pricing* is establishing how much the product will be sold for, and that is a function of top management and/or Marketing.

To blur for even a moment that definite and obvious distinction can lead to poor estimating. Thus it is wrong ever to allow the price that can be charged to influence the estimate. If the gap between estimated costs and the price that can be charged is too great, action may be needed to improve product design and/or production methods. That is well and good, and the estimate can then be revised downward to reflect the changes. But for a given design and for given production methods, the estimate must predict the costs without being influenced by the price that can be charged for the product.

A company can "design and produce to price," but it should *never* "estimate to price." Estimating deals with the realistic prediction of costs. If that prediction yields a price that will not be competitive, the company must (1) redesign the product, (2) improve manufacturing costs, (3) use a competitive price and accept a lower profit margin, or (4) do all three. Any mental confusion about the distinctive nature of estimating versus pricing can lead to coloring the estimate to reflect the feasible pricing range. In turn, this can lead to the failure to recognize the need for improvement and, of course, a completely unrealistic prediction of the product's cost. You *cannot* estimate to price.

Detailed versus Conceptual Estimating

Estimating has a tremendous range in practice and job content. At the easier end of the range is *detailed estimating*. It is not really easy, but the parameters with which we work in detailed estimating are more explicit and defined and the data available for our use are usually more numerous. In situations of detailed estimating it is usual to have:

Bills of material or enough data to develop usable bills of material
Parts prints or, at least, part descriptions to determine material costs

Parts routings or sufficient parts description detail to deduce the routing steps

Labor standards, of whatever source, or historical labor-cost data for calculating estimated labor costs

Typically, this kind of estimating occurs in foundries, machine shops, stamping shops, printing plants, box shops, die-casting and plastic-molding shops, and the like. Basically what is involved is estimating different models or varieties of existing products. In such shops Estimating occasionally finds itself involved in the other end of the range, i.e., conceptual estimating, because a radically new type of product or a very different type of existing product is being contemplated. In such manufacturing situations, however, the overwhelming majority of estimates is normally of the detailed type.

At the much more difficult end of the range of estimating is *conceptual estimating*. In situations of conceptual estimating the state of the art in the product is being extended. Thus, the product will have much greater accuracy, much higher productivity, and/or much tighter specifications, or the like. Since the product has never been made before, costs are involved for which there are no historical data or precedents. Not infrequently, however, parts, components, and subassemblies of the proposed item will be similar to those used in the past. Or certain subassemblies or groups of subassemblies used in past products will be used in the new product. In these cases, the estimator can use historical cost data on at least these segments of the new product's cost.

A form of conceptual estimating is often referred to as *parametric estimating* in articles in journals, transactions, and proceedings. The term normally applies to estimating in defense products such as airframes, aircraft engines, missiles, satellites, and advanced defense electronic products. It is rarely used in conceptual estimating of commercial products such as machine tools or heavy complex machinery and equipment for sale to manufacturing companies, institutions, or government use only indirectly for defense. In parametric estimating, a mathematical relationship is sought for past similar products or models, between actual manufacturing costs and the performance or physical characteristics of those past products or models. These relationships can be by total product, subassembly, or area of cost, e.g., labor hours, tooling, or engineering. A mathematical relationship or correlation, once identified, is used to estimate the cost of the new, advanced product with its advanced performance or physical characteristics.

Unfortunately a great many "games" are played in parametric estimating, in which deliberate estimating optimism is exercised by the producing company to obtain the contract. Worse, from the taxpayer's viewpoint, the buying agency in the government may participate in the game. This is candidly discussed in the article by L. J. Dumas listed in the Bibliography.

In this book, however, we are concerned with *realistic* predictions in estimating. The estimator in a profit-making company selling advanced products has the difficult chore of performing conceptual estimating to develop those realistic predictions. There is much to learn in this area, and the multiple-regression approach suggested for consideration and development in Chap. 7 is the beginning of parametric estimating in such a situation.

The Ten Most Common Estimating Problems

We can understand our estimating job better if we review some of the common problems faced by product-cost estimators. When we name these problems, we also specify many of the basic skills and disciplines a company needs to do an effective estimating job. The problems listed below are culled from my own observation and experience. You may have others to add. By reviewing these problems we can help define our most common difficulties. Only by defining the problems can we begin to approach solutions.

INADEQUATE DATA

Probably the single greatest problem faced by cost estimators is that of inadequate data upon which to develop their estimates. There are two major areas of such inadequate data:

Inadequate Product Data

One of the commonest cries in estimating circles is, "All they gave me was a sketch on a yellow pad, and they want an estimate." Frequently this situation is accompanied by the statement that "all we need is a ballpark figure." From past experience the estimator knows that within 2 weeks top management or marketing will have forgotten all about the ballpark statement and will treat the estimate as a precise prediction of future costs.

Frequently this situation is no one's fault, and there is no solution to the problem. For example, sometimes the prospective customer cannot or will not detail its specifications sufficiently to allow a proper estimate to be made. Marketing just cannot get a more complete set of specs. But since it is a service function, estimating cannot refuse to develop some kind of guesstimate. In other cases, when the information gaps are discovered and defined, it may be possible for Marketing to go back to the customer and obtain at least some of the needed information. Each situation is different.

However, where the necessary complete data cannot be obtained and an estimate must still be made, the estimator should do two things: (1) spell out on the estimate sheets the assumptions made in developing the estimate and (2) show on the estimate a projected plus or minus percentage by which the estimate could be in error if the assumptions used turn out to be wrong. Even though this plus or minus percentage is a guesstimate, it is a smart thing to include on the estimate.

Taking these two precautions puts on the record the inadequacy of the data with which the estimate had to be developed and reminds top management and Marketing of the potential error the estimate entails and the degree of risk that lies in the prospective job.

Inadequate Labor-Cost Data

Many estimators today are developing product-cost estimates without being given engineered labor time standards, i.e., labor standards developed by industrial engineers or time-study engineers, standard data from time studies, or preengineered times such as Methods-Time Measurement, Work Factor, or the like. As a result, these estimators have to use their own rough estimates of labor times (developed from their own experience or off the top of their head), supervisors' estimates, or Cost Accounting's history of past actual labor costs.

This usually represents a lack of important data, made even more important by the fact that in many industries labor hours and costs are the base to which we apply our manufacturing overheads. Thus, if a company has, say, a 300 percent manufacturing burden, any errors in labor-cost estimating resulting from inadequate labor standards are exaggerated by a factor of 3. Many managements fail to recognize this fact. For example, I've had clients in the electronics industry question whether labor standards are important in their situation because labor is only 15 to 20 percent of their cost of sale. But they then apply a 275 percent overhead factor on labor, and the possible errors in labor costs are multiplied by a factor of 2.75. True, there are some industries, like chemical plants, where labor costs are quite a small percent of the cost of sale and are not the basis for distributing overheads. Instead manufacturing overheads are spread over equipment hours or minutes. But in the common situation, where labor costs are the basis for overhead applications, the management should recognize the estimator's need for adequate labor time standards, and Estimating should constantly pressure management to obtain them.

Actually, in this matter of labor time standards, product-cost estimating rides the coattails of an even greater management need, that of achieving control of labor costs. The reality of this need can be demonstrated by three widely accepted principles of human effort.

• If you place a trained person on a job, a lathe operator at a lathe, or a typist at a typewriter, and just put them to work, you will get a 40 to 60 percent performance. To the uninitiated this seems like a very harsh statement, but to the manager experienced in work measurement it is simply a fact of life. I've had some managers mutter "How do you get 40 percent?" I've met a very few who believe that their supervision is so good that they average 85 percent, but it has to be questioned. The experienced industrial engineer or time-study engineer who has done any work measurement or work sampling will almost always agree with this observation. There are certain specific exceptions. For example, in a mechanized assembly line the speed of the line establishes the pace. But if the equipment does not control the output specifically in a management-controlled manner, you can safely bet that output will be 40 to 60 percent of what you would measure as 100 percent.

• If you use labor time standards to control labor costs, you must:

1. Establish the standard
2. Tell the operator(s) before they start the job what the standard is
3. Tell them how they did compared with the standard

Some clients say they have labor standards, but these standards turn out to be only cost accounting labor standards for standard costing. The people on the production floors do not know what the standards are or how their performance compares to them, and the company is not using these labor standards for labor cost control.

But if you do these three things, you are using measured day work (MDW) and can achieve 85 percent labor productivity. Of course, simply establishing the labor standards does not automatically raise productivity from an average of 50 to 85 percent. That takes intensive supervisory attention and work.

Some of the best-managed companies in the United States use MDW, recognizing that it is management's task to specify a fair day's work for a fairy day's pay. In my experience, however, MDW is greatly underused, and consequently thousands of companies lack meaningful labor-cost control.

• If you pay incentives, i.e., additional money for additional output, you can achieve a 120 to 125 percent labor performance. A company with an incentive plan that achieves an average performance to standard above 125 percent should begin to worry. Are labor standards being well enough maintained? Do these standards fail to reflect equipment, tooling, and methods improvements? The standards are simply too loose.

In some industries incentive plans are standard operating practice. An example is piecework in the ladies' garment industry, which could not exist without it. In many other industries where incentives are equally applicable they are rarely if ever used. In fact, I no longer recommend incentives to clients unless they have gone through 4 or 5 years of well-maintained MDW. Management history in this country is replete with the names of companies that allowed their labor-incentive standards to deteriorate to the point where the incentive plan actually restrained production output. Any management embarking on an incentive plan has to be psychologically prepared to insist on well-maintained labor standards.

Look again at the percentages stated above. If you could raise labor productivity from an average of 50 percent to 85 percent, think of the effect on product costs, particularly in these days of ever higher pay and fringe costs. This is precisely why some well-managed companies spend so much time and money in installing and maintaining MDW plans.

A company with good labor standards for labor-cost control can use them also for product-cost estimating, capital-investment decision making, make-or-buy studies, and a host of other areas for which management needs good labor cost data.

MASS QUOTING

In some estimating situations, frequently because of inadequate time and staff, a great deal of mass quoting is done. When a new request for quotations is received, the estimator examines it, identifies it as "like" a job done previously, and estimates it at the cost quoted on that past order. Or if the actual costs of that past job are known, they are used as an estimate for the new prospective order. I knew one electronics-component manufacturer who had 40 product lines with thousands of varieties in each product line and five product-cost estimators. There had to be a lot of mass quoting going on under those conditions!

This kind of product-cost estimating is a practical necessity in some companies and industries, but the nagging worry must be: Is this prospective job *really like* that old job? If it is, fine, but if the new job has tighter reliability specs or more costly requirements of any kind, the mass quote can yield a completely unrealistic prediction of costs on the new order.

When a company's estimating group is forced to engage in an appreciable amount of mass quoting, it behooves the management to make periodic and careful checks or follow-ups on actual costs versus estimated costs, at least on a sampling basis. It is all too likely, under these circumstances, that estimates on certain product lines will be outside acceptable ranges of error.

LACK OF COORDINATION BETWEEN ESTIMATING AND PRODUCTION

This problem occurs when Estimating predicts product costs based on the use of specific, high-production, low-cost plant equipment, but when the order is received it is actually made on less efficient machinery because the better equipment is tied up on other work. There goes the realistic prediction! Sometimes there is no way of avoiding this situation. If the second customer wants its units immediately, like the first customer whose job is on the faster machines, the new order must be run on the less efficient equipment.

From the estimator's standpoint, the ideal procedure is to use average capabilities, speeds, yields, and costs in developing the estimate. Thus, in a wiredrawing situation, if a given job can be run on machines drawing 300, 600, or 1000 feet per minute, the job would be estimated at

$$\frac{1900 \text{ ft/min}}{3} = 633 \text{ ft/min}$$

Then you are playing straight odds that the job, if received, will have a one-third chance of running on any one of the three machines. Often, however, the estimator must predict production costs under much more stringent conditions, e.g., a very competitive, low-margin situation in which the best available production costs must be used. If that is the case, the job will have to be run on the most productive equipment, namely, that used in developing the estimate.

A common cause of this kind of difficulty is not having cost estimates available to the production department. In many companies it is the policy not to have the shop aware of the costs estimated on a given job. Many reasons are advanced for this practice. Management may feel that if shop management sees the estimates they are more likely to live up to the estimated costs and be less efficient than they might otherwise be. Or management may be afraid that showing burden rates and markups to shop floor management will encourage them to project a final profit that is unrealistically higher than will probably be realized. All these feelings and fears, while unfortunate, are understandable but no reason for not telling shop management what machinery or equipment was used in the estimate.

If this is a problem, management should establish a system telling shop floor management and Production Planning and Control what equipment and production rates were used on a given job estimate. Estimating should take the lead in developing such a system and strongly advocate its installation.

IMPROPER OVERHEAD CHARGES

This problem is the result of the great errors many companies make in charging overhead to different product lines. Products not assigned their full share of overhead or not charged with the fixed costs they specifically incur are carried by the other products.

For example, a company has four product lines, A, B, C, and D, and because of the way overhead costs are assigned, product D is supported by products A, B, and C. A shocking percentage of people attending my seminars nod their heads affirmatively at that example. As a consultant and outsider, I have known production supervisors to point to a product and say, "Yes, we tape a five dollar bill to every one of them we ship." These people are telling you something that bears management investigation because frequently they are stating the realities of the situation.

All too often after hearing such a remark I have gone to the client and said in effect, "All I hear around here is that product D is a loser; why don't you determine how much it really costs?" The response is, "Why rock that boat? Look at that profit and loss. We're averaging out beautifully." This is bad and dangerous logic. Management is hard work with wearisome responsibility. Why work so hard to make a profit on A, B, and C and throw away some of the results on product D? Even though the profit is good now because D is a small part of total sales, that condition may change. For example, one float-valve manufacturer had a loser in a special stainless-steel valve. The company made a nice profit when the valve was only 10 percent of total sales, but when sales grew to 35 percent, it started to lose money. Since in the long run prices usually reflect costs, you can expect this to occur. If you are undercosting a product, you are more likely to underprice it, and when you do that your sales are likely to grow. There is no excuse for not knowing your product costs.

Proper fixed-cost allocation will be discussed in Chapter 3. It is sufficient here to say that Estimating should take the lead in correcting such a situation and should receive strong support from Marketing. In fact, Marketing itself should be taking the lead in correcting this problem, but unfortunately Marketing is frequently uninformed about product costing. Too often they blandly accept whatever they are told. As a result, it is Estimating that realizes the costing inequities between products. And they have to push for the corrective action.

CARELESS ESTIMATING

This problem may be either self-imposed or imposed on Estimating by other management areas, particularly Accounting.

Self-imposed carelessness occurs when Estimating fails to include specific costs that will be incurred in making the product(s) ordered. Such oversights can be of several types, e.g., the extra material costs of smaller-lot buying and the extra setup costs because of stretched-out customer delivery schedules, the extra costs of multiple shipments, extra scrap costs because of certain customer specification requirements, or just plain errors in calculations. The possibilities are legion.

It is easy to make mistakes in estimating. Many estimating situations deal with complex products and calculations; human beings are doing the estimating, even if aided in part by a computer. The estimator must reduce the seriousness and frequency of the errors and keep them at reasonable low levels. The need for correction will be clear from the findings of the estimating follow-up. The solution is better training and sterner discipline on the part of Estimating management.

More common than self-imposed carelessness are problems caused by other areas of management. We have already mentioned erroneous estimates that result from inadequate data. Another common example of erroneous estimating is related to improper overhead charges, reviewed above. The problem is caused by poor estimates developed when Estimating has only one overall manufacturing burden rate to use.

The classic situation is a job machine shop including a drill-press department, with many relatively inexpensive pedestal drill presses, and a planing department, with large expensive planing machines. In such a situation with only one overall shop burden rate, the job requiring a lot of drill-press work is being overcharged, and the job requiring a lot of planing work is being undercharged. In such a situation good estimating requires cost-center burden rates.

I ask each estimating seminar group, "How many of you have only one overall manufacturing burden percentage to apply to labor hours or machine hours to cover or include your manufacturing fixed costs?" Between 40 and 50 percent of the group reply that this is all they have to use. When I explain that they are undoubtedly undercosting some jobs and overcosting others, they invariably agree. It is amazing in this day and age that so many companies should have such a basic estimating and costing deficiency. The responsibility lies with the Controller and the cost accountants, who should be the source of the needed departmental burden rates.

Suppose the Controller's staff does not supply the needed burden breakdowns by cost center? Then Estimating should tell top management strongly and repeatedly that they need these data if management is to have better product-cost estimates. Estimating's needs may not be attended to immediately, but it is their responsibility to state them and keep repeating the statement until the data are made available. One unfortunate estimator told me that he has been after his company's Controller for 5 years to provide cost-center burden rates. The Controller admitted that Estimating needed them but never developed them. I could only commiserate with him and advise him not to give up.

This need for cost-center burden rates is not going to vanish, and the economic situation is exacerbating the problem. Investment costs per labor hour increase every half decade, and as a result, the future will see a need for even greater refinement in burden rates. For example, if you have a lathe department with a group of $50,000 lathes and two $300,000 numerically controlled lathes, Estimating needs burden rates by subcenter for that department, one subcenter being the numerically controlled lathes.

OPTIMISTIC ESTIMATING

In developing a realistic product cost estimate the estimator walks a razor's edge between optimism and playing it safe by overinsuring against possible unforeseen contingencies of cost. The difficulty concerning us here is optimism. Two examples come readily to mind.

• A company may have a standard scrap allowance of 3 percent built into its standard costs, but for the last 6 months actual scrap has been running 5 percent. In this situation any estimator who fails to build a 5 percent scrap allowance into the estimate is overoptimistic. Why should scrap on the prospective job be any different from what it has been for the last 6 months?

• Even more important is the situation in which labor is averaging, say, 70 percent efficiency to standard. The estimator is developing a bid that includes 70 standard hours of labor. Again, it is overoptimistic not to divide 70 standard hours by 0.70 efficiency and include in the estimate costs for 100 actual labor hours. Why should average labor efficiency be expected to be any better on this new job?

I have had estimators say, "But if I include 100 actual labor hours of cost, we won't get the job." This is again the classic confusion between estimating and pricing. To avoid it we must return to our basic definition of an estimate. It is a realistic prediction of *costs*. Pricing is up to top management and/or Marketing. Estimators must call the costs as they see them. Once management and Marketing know what the job will cost to do, it is their task to establish the price. If costs are too high for the price the market will bear, they must initiate the needed cost-improvement action.

In some enterprises, optimism is forced upon the estimators, who are not allowed to specify the costs as they see them because of pressures from other management areas. Since this problem is related to Estimating's position in the management hierarchy, it will be discussed in Chapter 2.

POOR ESTIMATING FOLLOW-UP

An estimate is a prediction of costs. Upon that estimate very often (and certainly in the long run) prices will be established and/or contract commitments made. It therefore behooves management to keep improving its predicting capability. Such improvements require a feedback, or correcting, loop in the flow of information within the company. This feedback loop is the follow-up or comparison of the costs actually incurred versus the costs originally estimated. Unfortunately many Estimating departments lack good follow-up information or the time to use it when it is available.

Estimating does the actual follow-up in three ways, depending on the business and product involved. (1) In a company making one-offs or big units in limited quantities the follow-up can be by unit number or job. Material, labor, scrap costs, etc., can be collected against individual units, and the follow-up can be by specific unit. Examples are power shovels, steam turbines, airframes, and aircraft engines. (2) Where production is by batch or lot, costs are incurred, collected, and recorded by production lot or batch. Therefore, follow-up of actual and estimated costs must be by cost per average unit in the production lot or batch. This situation is typified by manufacturing of metal parts, smaller machine tools, batch chemicals, electronic assemblies, and so forth. (3) In practice it may not be feasible to collect and calculate the actual cost of individual units or the cost per average unit of a given lot or batch in high-volume, continuous chemical manufacturing, oil refineries, soap plants, cement plants, etc. Here the follow-up record must be kept not by units but by cost account (labor, material, scrap, etc.).

Some manufacturing situations really are mixtures of the last two types. In making glass bottles, for example, the costs involved in making the glass itself, i.e., batch mixing and melting, are of the third type, and follow-up has to be by cost account. Actual forming of the bottles is done by production batches or runs, and follow-up can be achieved for each run made. A second example is in making corrugated boxes. In the corrugating department, realistic follow-up has to be by cost account, but printing, slotting, stapling, and taping are done by production run; for those costs, the cost follow-up can be in terms of the average cost of specific boxes.

How you are going to do estimating follow-up depends upon your product and your cost-recording procedures. That you should do it is self-evident. For adequate follow-up Estimating is most dependent on Accounting, the department that collects actual costs. Therefore Estimating must specify to Accounting what kind of follow-up information they need and in what format. These follow-up specifications may even have some effect on Accounting's chart of accounts.

The need for this tight interface between Estimating and Accounting may be why estimating follow-up is so poorly done today in so many companies. The harsh fact is that in many companies estimating follow-up is either almost nonexistent or inadequate. One of the reasons for this problem is that in many companies the staff and clerical help are not available to do the cost collection and compilation necessary for good estimating follow-up. The fact that Estimating and Accounting are using the computer more and more opens the way to adequate follow-up information and the ability to make successive improvements in estimating.

The basic need is for Estimating to recognize the importance of good follow-up data on actual costs versus estimated costs. As Santayana said, "Those who do not remember the past are condemned to relive it." First, you have to *know* the past.

IMPROPER PRICING

I have said repeatedly that estimating and pricing are two completely different matters, handled by two different areas of management. However Estimating is sometimes urged to develop a suggested or standard selling price. For this purpose profit markups must be used, and they are the problem at hand.

When Estimating does develop suggested or standard selling prices, they may be improper ones if an overall markup percentage is used. When the product being priced includes purchased components, one overall markup percentage applied equally to both purchased and manufactured parts can result in bad pricing.

Where used, these markup percentages, supplied to Estimating by top management, Marketing, and/or Finance, are averages developed to achieve desired relationships between cost of sales and the selling prices needed to meet the profit plan in the light of the sales forecast.

When a company makes a broad line of products having different ratios of purchased and manufactured parts, an average markup can be inadequate and even dangerous. Some products will be overpriced and some underpriced.

It is better for Estimating to be given a different profit markup for purchased parts and for parts manufactured by the company.

POOR ESTIMATOR SELECTION AND TRAINING

The quality of an estimate varies enormously with the industry, the company, and the time. In some industries, profit margins are liberal and the results of poor estimates have no drastic effect. In others, where margins are tight, good estimates and good costs are vital to a company's survival. Within a given industry, some companies enjoy a definite edge because of specific technological and/or managerial capability; for that company the quality and accuracy of the estimate is less vital than for some of its competitors. In booming times, when general prosperity can shield a company from the harmful effects of generally poor estimates, the quality of estimates and the capability of estimators can vary greatly without necessarily bringing serious harm to the company. In the longer run, however, the company's success will vary directly with the quality of the Estimating department and its individual estimators. It is obvious that a company needs to select and train its estimators carefully.

Consider one happy but dangerous extreme. A general manager of a steel casting company once said to me, "Our situation is ridiculous. We have more orders than we can handle. We keep getting requests for quotations and we don't want to ignore these customers. So we develop a quotation, throwing everything, including the kitchen sink, into the estimate. Then we double it, and send it off laughing. Then we get the order." Under such conditions it is easy to become fat, happy, and dumb. What's the need here for good cost data, for a capable, high-quality Estimating department? Anyone can estimate in such a situation. Actually, this is the time to worry and prepare. Inevitably the cycle reverses; it is during the good times, when we can afford them, that we should effect the improvements in cost data and estimator selection and capabilities.

Later we consider the selection and training of estimators. Here we merely want to emphasize that poor selection and training can be a real and costly source of trouble. Usually the difficulty can be traced to such practices as the following.

• Lack of management recognition of the sheer importance of estimating. This means that estimating is treated as a clerical, data-compiling function instead of the managerial judgment-rendering function needed for good product-cost predictions.

• Lack of sufficient training and particularly lack of exposure to other areas of management and management knowledge. For example, an experienced employee, e.g., a supervisor or tool and diemaker, may be taken into Estimating because of knowledge of the product and process, or similarly an engineer may be taken from Engineering into Estimating because of experience and technical knowledge and put right to work developing actual estimates without training or exposure, say, to cost accounting. Such people have no background with which to analyze or judge cost data provided them. This sink-or-swim procedure may be necessary for the moment, but should be changed as time goes on. Too many estimators are given too little time and help in developing their knowledge in fields they must master which are new to them because of their background.

MANAGEMENT INERTIA

Though this problem is listed last, it may well be the second most common difficulty faced by product-cost estimators after inadequate data. Certainly the estimator mentioned above, who has been trying for the last 5 years to obtain cost-center burden rates, is suffering from the results of management inertia.

Management's failure to provide the better data and training Estimating needs to keep improving their estimates is often the result of management's self-satisfaction: "We can't be doing too much too wrong; look at that bottom line." "Our costing and estimating are working fine. The profit and loss statement proves it." This is a non sequitur, an incorrect inference from the profit and loss facts. It is fine to be doing well in terms of profit, but the real question is what must be done to make an even better estimating job possible. Then the profit will be even better, and the company will be better prepared for the poorer times that cyclically occur.

Unfortunately management inertia is often the result of Estimating's failure to communicate strongly and repeatedly:

1. The need for additional data
2. The kind of Accounting action needed for better estimating follow-up
3. Suitable time frames from Marketing and Marketing's prospects
4. Other areas needing improvement

All in all, the above ten are an imposing set of problems. They have been discussed (1) to introduce the basic understandings and disciplines needed for good product-cost estimating, (2) to strike some familiar chords, reminding readers that they are not alone, and (3) to reassure them that this book deals with real and practical matters of estimating.

Factors Increasing the Difficulty and Importance of Estimating

Much evidence supports the observation that product-cost estimating is simultaneously growing more difficult and becoming better recognized for its importance. More and more companies are establishing Estimating departments. There is greater interest in management seminars on estimating, and there are more executive recruiting advertisements for Estimators. Let us look at factors making estimating more difficult. Some are quite obvious; others are more subtle.

WIDER PRODUCT LINES

As product lines proliferate, the task of estimating becomes more difficult. There are more parts, more variations of given parts, and more combinations and permutations to be handled. Certainly it is harder to estimate for eight product lines than for six.

MORE MODELS WITHIN PRODUCT LINES

Many companies have an increasing array of models within product lines, and as a result, Estimating has more variety to deal with. It is here that the problem of *mass quoting* can often be seen. A glass-bottle company can make hundreds of varieties of 16-ounce flint glass bottles, and the problem is similar in manufacturing electronic components and electrical controls. The more models, the more the care Estimating must exercise to ensure that they are recognizing cost differences between models.

TIGHTER ENGINEERING AND QUALITY SPECIFICATIONS

Modern industry struggles for more and more control over the manufacturing process and is less and less satisfied with empirical and trial-and-error approaches. We want to know why things happen so that we can exercise better control over our production processes. As a result, our engineering and quality specifications become more and more precise, and more demanding, and the higher the standards, the greater the costs. In a given company engineering and/or quality standards may vary markedly between product lines and even between product models. This makes the estimating harder because Estimating must differentiate between the engineering and production costs of these products with different engineering and quality specifications.

INCREASING COMPANY DIVERSIFICATION

This is less common, but some estimators find themselves having to develop product-cost estimates in product fields that are completely new to them. This situation arises when a company buys another company manufacturing a different product and estimating inadequately. Then Estimating must take over the estimating for the acquired company. As companies diversify, this problem may become more common.

CHANGING CAPACITY UTILIZATION

Since inflation keeps increasing the cost of production equipment, management becomes more intent on improving equipment utilization. As capacity utilization increases, the smarter management wants to devote that equipment to work that yields the best contribution margin or the best profit margin, thus improving the return on investment. Since management does not want that equipment tied up in lower-margin work, it requires increasing definition and of certainty the cost differences between jobs. Decision making on the basis of averages becomes unacceptable.

The obverse can also be observed. When times are booming, the backlog is great, and the profits are up, concern with better estimating may languish. Obviously management affected by the "fat cat" syndrome is not smart management.

Capacity utilization generally varies directly with the business cycle. In good times management seeks to use the company's available capacity on the most profitable work, which can be recognized only by good estimating. In slow times, when utilization is down, the pressure is to get enough work in the house to keep production above the breakeven level at least. On the cycle's downside good estimating is needed as the company fights to keep its skills and its capacity reasonably intact until the cycle reverses. Under these harsh conditions, estimating is more difficult both because it has to be closer and because it frequently is called upon for allied but different outside work needed to keep the operation going. Clear examples of this have been seen among steam turbine manufacturers in the early 1980s.

In the United States we have had more ups than downs in the past three decades, but cycles do occur, and when they do, the demands for a better estimating performance tend to increase.

GREATER QUALITY AND PRICE COMPETITION

Other industrialized nations are imposing increasing quality and price competition on the United States, and in many American industries there is already stern quality and price competition. Under such competition the companies with the better cost estimating will do better than the others. Of course, good cost estimating is not the only answer to competition: marketing, product design, manufacturing costs, and quality control are all vital. But these factors must be supported by good product-cost estimating.

One of the worst types of competition you can have is the company so ignorant of its costs and so poor at estimating that it quotes completely unrealistic prices. It issues quotes that cannot possibly cover costs, much less yield an adequate profit. What can you do with such a competitor? When the competitor is trying to "buy the business," your management can either match its bid (if you want or need the order badly enough) or let the job go. In some cases, however, the competitor is simply ignorant of its real costs. In that event, Marketing might go to the customer and explain that you know your costs; the competitor cannot possibly make the item profitably at the price quoted and has to lose money on the job. Suggest that at that price the customer may well have quality or delivery problems when and if the competing company discovers its errors. This can sound like sour grapes to the customer, but your advice may be helpful and work to your advantage in the future. The saving grace for many companies faced with such underinformed competition is having other product areas where the two companies do not compete. They can hold out until the competitor either goes broke or recognizes its errors.

One practicing estimator told me that although he and his one competitor are very close in their quotations for the first 9 months of every year, every fall the competitor goes berserk, issuing quotes that cannot possibly yield a profit. Apparently the competitor reasons that by fall the year's worth of overhead has been covered and marginal pricing will be safe for the last quarter. That is tough competition! Smart buyers would try to concentrate orders with that competitor in the fall.

INCREASING CAPITAL INVESTMENT

Since capital investment in manufacturing increases every 5 years, better estimating becomes more and more important. But better estimating is more difficult. When you are estimating worker and machine minutes and hours on a $30,000 engine lathe, you can afford to be less accurate than when you are estimating minutes and hours on a $300,000 numerically controlled lathe. Increasing capital investment has an even more subtle effect on the ramifications of estimating. In the future, the estimator must have closer liaison with other management areas. It is the wise estimator who does not always accept the first program for the numerically controlled lathe offered by the programmer but works with the programmer to improve the efficiency of the program finally used. A good estimator will check out the sequence of cuts, the speeds and feeds used, and so forth. Improvements and a lower estimated cost may be possible. A good estimator does this with manually operated engine lathes, but it becomes even more important with the numerically controlled lathe.

BETTER, MORE WIDELY DISSEMINATED MANAGEMENT TECHNIQUES

As in engineering, medicine, and many other fields, more techniques are becoming available for use by manufacturing management. The growth appears to be exponential. Estimating practices of 20 or even 10 years ago are unacceptable today. This means that the estimator has to know more and is under more pressure to keep learning more. Not only are there more and better techniques available, but there are more avenues to learning—books, professional journals, continuing education programs, university courses, and professional seminars.

In fact, the problem today becomes one of selection. Estimating, like other management areas, is the art of the possible; under present conditions many of us cannot use all the techniques available. Practicing estimators thus must not only know about new techniques but must also recognize which techniques have practical use for them. As one estimator put it, "I have a master's in industrial engineering from Stanford. I know all about mathematical modeling and operations research. But hell, out here in the real world you can't even get the data." Still, he is well off to know the techniques since, knowing them, he is more likely to recognize a future situation where they can be used. Like any good manager, an estimator has to be a student and never stop learning.

GREATER MANAGEMENT SOPHISTICATION

Allied to this growth of management techniques is the greater sophistication of senior managers. More and more of them know about new managerial techniques through education or experience or both. This means that their demands on Estimating and other management areas will be more sophisticated.

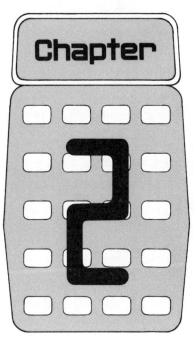

Estimating's Personnel and Place in the Organization

This chapter deals with the people in Estimating. It discusses where the function of Estimating can be found in various organizations and the pros and cons of each such location. The qualities needed by a good estimator are specified, and the selection and training of estimators is considered.

The Need for a Separate Estimating Function

When a company is small, the task of estimating is rarely identified as a separate function in the hierarchy of management. In a very small company, the owner or operator alone makes the best possible estimate of the costs involved in the job, but normally in a small company estimating is handled in one of two ways.

• The various departments submit estimates of the job's costs; Production estimates labor costs and tool costs if applicable; Purchasing estimates material and purchased-component costs; Engineering estimates engineering costs; and Accounting provides the overhead rates to be applied. Then one department, frequently Accounting, adds everything together to develop the final estimate. There is no specific Estimating department that is responsible for the quality of the final estimate;

instead many different departments are ostensibly responsible for their individual cost inputs.

• Estimating is a part-time chore in one department. Production management or Cost Accounting may perform estimating as part of their other work. They collect data from other departments such as Purchasing and Engineering. This arrangement differs from that above in that the task of estimating is recognized and partially separated, though it is not the exclusive task of the department performing it.

Either of these methods is a realistic approach when a company is small. The chief difficulty is that all too often it is continued into levels of company growth and size after it is no longer adequate. It is startling to see companies with $50 million in sales volume just beginning to establish a separate and independent Estimating department.

Obviously, the quality of an estimate has too great an effect on the company's operating performance for it to be slighted or handled in a cursory manner as just another chore. Any top management that recognizes and accepts the importance of good estimates must make Estimating a separate, independent function as soon as the company's size indicates the need for such action.

The level of company size that requires Estimating to become a separate department varies according to the technical nature and complexity of the product or products. Thus, with simple products consisting of one or few parts estimating in many instances can be found being done by Cost Accounting, even though the company may be fairly large (the glass-bottle industry is an example). Not infrequently in these situations estimating is really only a clerical, cost-compiling function, with little voice in, and less influence on, management decision making. Rarely in such situations will you find Estimating involved in technological changes and capital-investment decision making or in decisions about product contraction and expansion.

Even in this type of simpler product enterprise, however, estimating problems can all too easily arise as the company grows. For example, at a two-plant level the home plant's Cost Accounting does the estimating. If there are four, five, or more plants and they still do the estimating centrally, they may fail to recognize the cost differences that inevitably arise between plants. As a result, they estimate on averages and no longer turn out realistic predictions of product costs.

In contrast, the more complex the product, the more likely that the need for a separate, independent Estimating department will be recognized earlier, when the company is not as large as one producing a simpler product. This is true because with a more complex product line estimating is more intricate, it is easier to make errors, and the errors are usually much more expensive and more obvious. Because of the more difficult estimating situation management is forced to realize the need for high-quality estimating at an earlier stage of company growth.

In this book we assume that senior management has recognized or is now recognizing (1) that the company needs good estimating and (2) that this calls for a separate Estimating department to provide an *independent, realistic prediction* of product costs. Thus, although Estimating may collect many of its cost parameters from Production, Purchasing, Accounting, and/or Engineering, only Estimating is assigned responsibility for developing the final realistic product-cost estimate. With that responsibility they have not only the right but also the duty of questioning matters like the following:

Unrealistic labor standards from Production

Overoptimistic scrap percentages provided by Production

Proposed purchased-component or tooling costs that could be bettered if Purchasing conducted further source searching

Inadequate and/or illogical burden rates from Accounting

The list could be extended, but the point is plain. Estimating should be a separate function with a clearly assigned charge to develop a responsible, independent, realistic estimate that represents a usable prediction of the product's cost.

The Estimating Department's Location in the Organization

In Chapter 1 an estimate was defined as an *independent, realistic prediction* of what it will cost to make the product. Both the key words *independent* and *realistic* are invariably and often drastically affected by Estimating's placement in the company's organization.

Estimating's location and rank in the management hierarchy usually depend on the following two factors.

The Type of Product

The greater the product complexity and the more highly it is engineered, the higher Estimating's position will tend to be. These are the same product conditions that cause a company to recognize the need for a separate Estimating department early in its growth. The more complex and more difficult the estimating task, the higher its rank will usually be and the more likely it is to participate in management decision making on such matters as methods and technology changes and product contraction and expansion.

Management Awareness

The more the company's executive management is sensitive to the effect of estimating and estimating follow-up on the profit and loss results, the higher Estimating's position will tend to be. The more management demands of Estimating in the way of refined estimating procedures, higher orders of estimating accuracy, and better controls over performance to estimate, the higher the slot Estimating tends to occupy in the hierarchy.

If management awareness is not particularly keen, Estimating's position may be the chance result of historical precedent and the organization's structure—the workings of company politics, in other words. If Estimating is under another function with a dynamic and politically effective executive, its recognition, importance, and participation will be enhanced. Of course, the reverse can also be true.

Most Estimating departments will be found in one of the five positions shown in Figure 2-1.

ACTUAL LOCATION

From responses of those attending my estimating seminar I have compiled the following estimate:

Location	Percent
Production	50
Engineering	25
Finance or accounting	15
Marketing	5
Chief executive officer	5

(I have heard a few people mutter, "Or all of the above.")

As expected, there are great differences in practice with different cost segments of the estimate. For example, in estimating situations involving a great amount of engineering time, product-cost estimators may be responsible for estimating manufacturing costs while engineering estimators are responsible for estimating engineering costs. This may be a realistic approach in some companies, but it is doubtful whether responsibility for the final estimate should be subdivided. It may be satisfactory if product-cost Estimating is held responsible only for manufacturing-cost predictions and not for Engineering's estimates

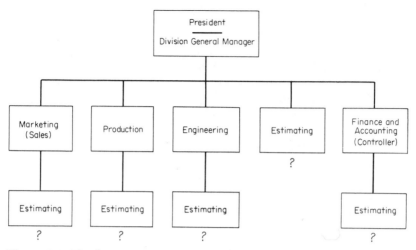

Figure 2-1 The five common positions of Estimating in company organization.

of engineering costs, but it leaves senior management with a less clean-cut assignment of responsibility for the total cost estimate.

Another variant, almost always less desirable, is where product-cost Estimating is responsible for estimating manufacturing costs and Purchasing is responsible for estimating the costs of purchased components. Thus, the total estimating responsibility is split. I see no advantage in such subdivision of the total estimating task. Purchasing can certainly be the source of costs for purchased components going into the final product, but Estimating should not be relieved of the responsibility for the quality of the final total estimate. Estimating should be responsible for questioning Purchasing's vendor quotations if they seem too high and not simply accepting whatever Purchasing comes up with.

Estimating's place in the organization is so important and has such a great effect on the independence of the estimate that it is worth reviewing what can happen to Estimating in each possible location.

Under Marketing

Probably the most dangerous area for Estimating is under Marketing. This harsh judgment is made as the result not only of my own 35 years of consulting experience but also of the experience of a great many practicing estimators.

The reason is obvious. When Estimating is under Marketing, there is a real danger that it will be under pressure to color or modify its cost predictions to conform more closely to the prices that Marketing believes the market will bear. This is not the road to independent, realistic estimating.

In all fairness, it should be added that two or three seminar participants have told me of being under Marketing and being free of all such pressures, but they are a small minority. I have had many, many more practicing Estimators tell me that they were under Marketing and had experienced such pressures to revise the estimates they judged to be realistic.

Theoretically, Marketing should be objective and of all the management functions the one most anxious to know the realistically predicted cost of the products they sell. But we manage in the real world with real people, and the pressures on Marketing to achieve increasing volume, for example, can in real life lead them to apply these pressures on Estimating.

Under Production

In most companies Estimating will be found under Production, the best and most practical place for it in many companies. We should recognize, however, that again there is danger of pressure being brought to modify estimates. Two such kinds of pressure are 180 degrees apart.

The pressure on Estimating that I have observed occurring most frequently is urging Estimating to use unrealistic cost standards that have been issued but never achieved. Examples resulting from overoptimistic estimates of labor productivity or scrap allowances were mentioned in Chapter 1.

At the other end of the spectrum another type of pressure may be exerted by Production management. As one practicing estimator put it, "My Production Manager studied the estimate and said that this was going to be a very tough job so I had better add another 10 percent insurance."

Again we have the conflict between theory and practice. Theoretically neither kind of pressure should occur. But they do and are a danger to which we must be sensitive. These pressures can affect the independence realistic estimates require.

Under Engineering

The more highly technical the product, the more likely it is that Estimating will be found under Engineering. Nevertheless, in many companies producing a highly technical product Estimating is under Production. In some such cases, as already noted, the responsibility for estimating manufacturing costs may come under product-cost Estimating and that for estimating engineering costs under engineering Estimating.

When product-cost estimating is under Engineering, one danger is particularly observable in electronics companies and in companies producing machinery with many parts and many levels of subassemblies. Engineering lays out the bills of material and specifies the sequence of subassemblies and their composition, i.e., the parts going into each subassembly. But in actually producing the product Production follows a different subassembly sequence or assembles specific parts into subassemblies differing from those specified by Engineering. How Production actually makes the product differs from the way Engineering projected it. The costs will therefore be different. Estimating knows from experience that a realistic estimate must reflect the costs predicated on the way it will actually be done. As long as they are under Engineering, however, they must estimate according to how Engineering sees it being done. The result is an unrealistic estimate of product costs. This is not a hypothetical situation; it actually occurs.

Under the Chief Executive Officer

When Estimating reports directly to the top executive who bears the responsibility for the profit and loss statement, Estimating will probably be armored against the kinds of pressures we have been reviewing. However, in the organization patterns of most manufacturing organizations it is usually impractical to have Estimating so positioned, because the chief executive officer's span of control can rapidly become too broad.

In the past Estimating often reported directly to the chief executive officer in such manufacturing milieus as airframes and aerospace work, where contracts involve hundreds of millions of dollars and the quality of the estimating job is vital to the company's continued existence. Recently, however, two airframe divisions in different corporations have shifted Estimating from under the chief executive officer to under Production. In substance, it is the span-of-control problem that frequently makes it unrealistic for Estimating to report directly to the chief executive officer.

Under Finance and Accounting

This position for Estimating has been saved until last because it can be either very good (possibly the best) or very bad. It all depends on the quality of the senior executives of Finance and Accounting.

Many different kinds of organization exist for Finance and Accounting. Large corporations, for example, have a financial executive, frequently called a financial vice president, responsible for dealing with bank loans, stock and bond issues, cash flows, cash usage, etc. A separate position may be an executive, called a Controller, responsible for all accounting, including cost accounting, and for tax matters. Frequently, in such situations, the Controller (divisional Controllers if they exist) reports to the corporate vice president of finance, or director of finance, or whatever the title. In large multidivisional companies, a vice president finance may be at the corporate level, but divisional Controllers commonly report to the divisional general managers, who have profit and loss responsibility for their division. This is the preferred way because a division general manager with profit and loss responsibility should also have the controller for the division under him, even though that Controller follows corporate policy and guidelines in constructing the division's financial statements.

In the medium-sized and small company a much more uniform pattern has developed. The corporation usually has a Controller, who is responsible for all financial and accounting matters. Let us consider such a situation.

Properly conceived and staffed, the controllership is a top management position. It reports directly to the chief executive officer and is at an equal organizational level with the heads of Marketing, Engineering, Production, etc. Although the Controller may have risen from the ranks in Accounting, this executive has trained and developed into a broad-gauge businessperson familiar with the practical realities of Marketing, Engineering, and Production. If the controller is such an executive, who may even be in serious competition for the position of chief executive officer, Estimating can be well placed under Finance. So placed, Estimating can be well buttressed against influences from other management areas, and the probability of independent, realistic predictions of product costs will be increased.

However, the harsh realities are such that not all controllers are executives of this type. This evaluation is not only the result of my own experience in management consulting: too many experienced estimators make the same observation. In some companies the controller is actually the chief accountant, who, like the people under him, may not realize, appreciate, or react to the realities and needs of the other management areas. If they do not recognize and understand those requirements, and if he directs the estimating function, the product-cost estimates may well be nothing but compendiums and projections of past costs and not the realistic predictions needed. Estimating placed under such a controller is less likely to have the knowledge to question bad production standards, dubious engineering estimates, or poorly shopped purchased-material cost projections. Under such conditions, Estimating is poorly placed under Finance and Accounting. It would be better placed under Production.

These are unpleasant statements but need saying. The controllership is too essential to be inadequately staffed. Typically:

Marketing needs good product costs to develop good product prices.

Engineering needs good cost data to budget and control performance on its many projects.

Production needs prompt and accurate reporting of actual costs for effective cost control.

Estimating needs good actual past costs for estimating follow-up.

Estimating and the company need logical overhead or burden rates for realistic product costing.

In summary, it is impossible to state any one best place in the organization for Estimating to be. It all depends on the talent aboard and the given organization's political realities. Perhaps the single most common place is shown in Figure 2-2. The one thing you can say is that senior management should always be sensitive to the real-life pressures that can be brought to bear on Estimating and should see to it that Estimating is placed where it can and does develop product-cost predictions that are independent and realistic.

Qualities of a Good Estimator

If management is to have quality estimates available for sensible product pricing, Estimating must have qualified estimating personnel.

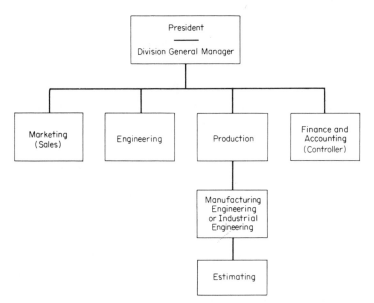

Figure 2-2 The most common position for Estimating.

What qualities identify a competent estimator? They include the necessary knowledge of seven special areas and three personal traits.

AREAS OF EXPERTISE

The Product

Obviously, the estimator must know the product lines intimately—not only the present lines but modifications to the existing products and, on occasion, completely new products. To have a detailed knowledge of the product, the estimator should be familiar with design parameters and how the product works. In the case of electronic products, for example, the estimator should have a good knowledge of electronics, of electronic components and what they do, of circuits and subassemblies, and how they work together to accomplish the product's purpose, and of testing procedures and requirements. Estimators do not have to be design engineers but the more they know of the product, the fuller and better the estimate.

This requirement helps explain why so many estimators have been recruited from the production floor or from Engineering. The more technical the product, the greater the need for detailed knowledge of it. But even for simple products, the estimator's knowledge must be complete.

The Process

Just as clearly, the estimator must know the production processes. For metal and machined products this means machine tools and machining; if sheet-metal work and/or welding is involved, it means knowing sheet-metal work and/or welding is involved, it means knowing sheet-metal working equipment and/or welding processes and equipment. For electronics it means printed-circuit-board manufacturing processes, harness and circuitry production, and testing methods. For chemicals it means the chemical plant's processes. The more estimators know of the process, the better they can choose the most efficient methods and manufacturing steps to build into the estimate and the more helpful they can be to Manufacturing, Engineering, and/or Production management in selecting the most cost-effective production processes.

The Materials

Good estimators have thorough knowledge of the material or materials used in the final product. This knowledge extends also to purchased components going into the final product. Such knowledge of material is essential if the Estimator is to:

• Specify and estimate the proper material costs.

• Detect possibilities of material substitutions when Design Engineering specifies the material, allowing multiple sourcing and/or lower-cost material or components.

• Select the correct source or sources of purchased material and/or purchase components. The final decision on material sources should be Purchasing's, governed by Design Engineering's specifications, but estimators themselves must often select possible suppliers and their price lists because there is not enough time to go through Purchasing in developing the estimate.

• Be able to evaluate the quality of predicted material costs developed by Purchasing. Many a good estimator has detected and corrected vendor proposals approved by Purchasing but evaluated as too high by Estimating.

The Routing

Routing means specifying the sequential operations needed to make a given part and/or the final product. These sequential operations are listed on route sheets, process sheets, operation sheets—whatever they are called in the specific plant. In some situations route sheets are not needed because the sequence of manufacturing steps is inherent in the process. After a given operation the material can go only to one subsequent operation, e.g., glass and glass-bottle manufacturing, most chemical plants, and oil refineries. In such situations, while the estimator must know the process, routing is not required.

In the manufacture of many products, however, operational routing is an integral part of the product-cost estimate. Some estimators are given the parts and/or product routing by Industrial Engineering or Manufacturing Engineering, but even in that happy case the estimator should be able to evaluate the quality of the routings provided. With the capability to evaluate, estimators may be able to suggest alternate routing yielding lower costs. Certainly an estimator should not have to accept on faith any routing that happens to be supplied.

In other manufacturing and estimating situations routings are not supplied or time is too short to obtain routing data from other departments. Many an estimator has to work from sketches or blueprints saying only, "Machine complete" or "Machine as per blueprint." No routing steps are provided. Under these conditions, to develop a good estimate the estimator obviously has to know how to route the part, the subassembly, or the final product.

Labor Standards

Labor costs are an integral part of a product-cost estimate. What percentage labor costs represent of the total cost of sale varies, of course, with the product. But in most estimating situations, labor costs are an important part of total product costs. It was pointed out earlier that their importance is increased by the fact that in many product-costing situations they are the base upon which manufacturing overhead costs are applied. The product-cost estimator had better be knowledgeable in the area of labor standards.

In some companies estimators are provided with labor standards, by operation, time study, or industrial engineering function. But again, as in the case of routings, estimators should be able to evaluate the quality and accuracy of the labor standards provided. Since estimators are responsible for a realistic estimate, they must be able to tell whether provided labor standards are too optimistic or, at the other end of the scale, have too much insurance built into them.

In the less happy situation estimators are given no labor standards by another department and must estimate setup and/or cycle time standards for every operation required by the routing. This situation is so common that I judge it to be the case in most estimating situations—and for a very good reason. In this country since the 1950s there has been a clear decline in the number of companies who develop and apply labor standards for labor-cost control. Incentive plans have decreased markedly. Too few companies have yet installed measured day work (MDW) plans. In new companies and in new industries, like electronics, for example, companies that have developed and applied labor standards for labor-cost controls are in the minority and probably represent far less than 50 percent of the total. As labor costs and the costs of fringe benefits increase and the need for good labor-cost control becomes more apparent and vital, this may well change. But today many estimators work without being provided with engineered labor standards. These estimators must estimate labor times, which means having a good grasp of labor standards.

Cost Accounting

Another area in which it behooves estimators to have a good working knowledge is cost accounting. In many if not most estimating situations, estimators use data supplied by Cost Accounting. Admittedly, the degree of need and use varies with the product and with the specific company situation, but all estimators have some need for such data supplied by Accounting. It is vital that estimators be able to evaluate the logic, completeness, and accuracy of those data. Otherwise the data must be taken on faith, and that may lead to a bad estimate.

Consider only one cost area, manufacturing burden rates, which will be discussed in more detail in a later chapter and another context. All estimates must include manufacturing fixed costs if they are to be complete. The burden rates that allow those costs to be included are invariably supplied by Accounting and in most cases by the cost accounting group in the Accounting department. In some situations, as we shall see later, it is Accounting that actually applies these burden rates in developing the final estimate. In any case, no matter who actually applies the burden rates, they must be logically calculated and applied. Since Estimating is or should be responsible for the quality of the final cost prediction, Estimating should approve of the burden rates used, how they are developed, and how they are applied. A company can have accurate, complete bills of material, excellent operational routings, and good labor standards and still not have good product-cost estimates if the burden rates are not logically and properly calculated and applied.

What I am advocating here may well conflict with the understanding of many experienced estimators and cost accountants. Many estimators believe that Estimating has nothing to do with allocations of overhead and the burden rates used to achieve such allocations. Estimating may apply the burden rates, but the estimating staff looks to Cost Accounting to calculate those rates and to furnish them to Estimating. Cost Accounting holds equally firmly to the view that such burden-rate development is strictly their province. Basically, I agree. It is Cost Accounting's task to calculate burden rates, but Estimating is responsible for realistic predicted product costs. Therefore, they have not only the right but also the duty to question the burden rates they are given to use if they are convinced that such rates are illogical, are too gross as averages to be usable, and are leading to undercosting some product and overcosting others. Many estimators are given just such burden rates to use. Inadequate rates should not be accepted supinely or used blindly. Good estimators know enough to question them and to keep questioning them until the rates make sense and allow realistic product pricing. More will be said on this important matter in a later chapter.

Marketing

Ideally, good estimators should be given some exposure to the realities of Sales and Marketing and their field work, but in real life this is rarely done. Only a small minority of estimators are given the opportunity to go out into the marketplace with Sales and Marketing personnel and to witness at first hand some of the customer uncertainties, competitive pressures, and psychological vagaries that represent the realities of Marketing. Since most estimators came from the ranks of Production, Engineering, or Accounting, they would benefit from the opportunity to see such conditions of work. Such exposure can increase their understanding and empathy with Marketing, one of the functions so closely affected by their product-cost estimates. What they see and learn out there in the field should not make their estimates less realistic, of course, but must help them realize how important a quick response from Estimating is to Marketing. Also the company benefits by having more knowing estimators and perhaps eventually even a well-qualified addition to Marketing and Sales.

These seven areas of knowledge represent a broad scope and demand a wide grasp of the manufacturing and business management environment. Where do you obtain people with such knowledge? Obviously, they are seldom, if ever, found fully developed. Most companies have to train their own estimators on the job, but before we consider the source of good estimators, let me mention three personal characteristics I always try to look for in choosing an estimator.

PERSONAL CHARACTERISTICS

Intelligence

I have always believed that a good estimator must have above-average intelligence. Too much rides on the result of estimates for a company to afford low intelligence in Estimating. Along with good intelligence an estimator should have an open mind, i.e., unwillingness to be bound by preconceptions or past practices and a need to inquire and search out new ways and new ideas.

Curiosity

Good estimators have a curious mind. They want to know more. They ask questions. They delve into areas of knowledge that are allied with, if not directly involved in, their estimating job.

For example, many a good estimator has been recruited from the ranks of production supervision. In metalworking companies you will find very able estimators who were first qualified tool and die makers. Their abilities were recognized and they stood out from their peers because they were curious. They evinced an interest in other areas of the business; they wanted to know about other parts of the enterprise. Standing out among their peers, they were recruited for estimating, which needs a mind with broad knowledge.

Logical Mind

Above all, good estimators must be logical. They must have a logical mind, which is not always a concomitant of high intelligence. For example, people with intelligence and creative ability are not always logical in their thinking. Often in working on a consulting assignment on product estimating and costing I have found myself sitting back and thinking, "This is really an exercise in logic." In working with client personnel and in training clients' estimators, it is necessary to be alert to their consistency of reasoning. Anyone doing estimating must be sure that the final estimate is a cost structure that is logical, that no one in Management or Marketing can find errors in its logic. The point is of particular pertinence in the allocation of overheads or fixed costs.

In real life estimating entails many value judgments. Their quality will vary directly with the ability of the estimator to think logically. The fact that estimating is so affected by value judgments is repeatedly impressed upon me at estimating seminars. For example, I give each session a case problem in estimating in which we consider the question of make-or-buy (it is basically the problem given in Chapter 8). The whole group at the seminar, say 24 people, is broken up into 12 pairs, and each pair calculates the estimated cost of the item. Usually they have 12 different answers, some close and others 200 percent or more apart. Some of the spread is due to the way overheads are applied, but even groups that apply overheads in exactly the same way have widely different answers. It happens at seminar after seminar, although most of the people attending are professional estimators. Even these professionals are applying different value judgments in arriving at their individual estimates. And again, good value judgments require clear, logical thinking.

Psychological Testing   These three personal traits are not always easy to identify. You can frequently detect a curious mind and wide-ranging mental interests by studying the applicant's background, interests, and avocations. Good intelligence can sometimes be reasonably evaluated by the applicant's past accomplishments and progress through formal and on-the-job training. But what may be the most important of the three, a logical mind, can be much more difficult to evaluate. Here industrial psychological testing can be of great assistance.

Today there appears to be a trend away from the use of such testing. Many managers have always been suspicious of its real value. Of more importance are the pressures some managements believe exist today under fair employment and equal opportunity guidelines established by the government. Some people at my seminars have told me that such psychological testing is no longer permitted in their companies. The fact remains, however, that it is possible to give tests for certain abilities if the critical content or skills required by the special job have been properly validated. Seminar participants who do estimating for federal establishments have told me that their group does such testing in selecting among applicants for estimating work.

One of the areas management needs the most help in is in selecting personnel. When you have two or three possibilities for one estimating opening, you need all the help you can get in selecting the applicant who offers the best likelihood of succeeding at the job. In Estimating the person selected is going to get a great deal of training on the job. Since that is expensive training, you want to increase the odds that such training will be effective and not wasted. Why not have the applicants take the Watson-Glaser Critical Thinking Appraisal Test and/or the Miller Analogies Test? The results can be of help in evaluating which applicant has the most logical mind.

Sources of Good Estimators

Where do you find good estimators? In rare instances the company may be able to find them from sources outside the company. Sometimes they are found through advertising or agencies. However, these ways are probably the least successful. In other instances, a competitor or a company in an allied field may be moving or cutting back, and one or more estimators will be looking for openings. For example, our airframes companies have even recruited abroad for estimators. Sometimes a good estimator in another company can be attracted by higher pay and/or benefits or greater opportunity for advancement. (Some people call it pirating; others call it competition.)

Usually however, the most productive source of good estimating talent lies within the company. For most companies it is generally most practical to identify on-board talent that can be recruited into Estimating. Such people may need further training to be developed into qualified estimators, but they know the product and/or process and management knows them. As such, they offer the most likely source.

At the estimating seminar I ask participants what management area represented their main background before getting into Estimating. Their responses are roughly as follows:

Main background	Percent of total
Engineering (industrial, design, and tool)	40
Production	30
Accounting	20
Others (material, marketing, etc.)	10

In terms of the nature of the product lines involved, from the most highly technical to the simplest, the sources of estimators appear to fall into three categories.

Highly Technical and Complex

Most are selected from Engineering, particularly Design Engineering, fewer from Production, rarely from Accounting. Examples of this type of product area are aerospace products, airframes, aircraft engines, and complex electronic system products.

Medium Complexity and Technology

Most are selected from Production, fewer from Engineering, usually industrial or tool engineering, and fewest from Accounting. Examples of this type of product area are heavy and light metalworking for metal products, construction, etc.

Simpler Products

Most seem to be selected from Accounting, fewer from Production, and only a very few from Engineering. Examples are food products, toiletries, soaps, simpler chemical products, glass bottles, and the like.

There are, of course, thousands of exceptions to these general classifications. I have seen accountants doing very well in estimating situations involving complex product manufacture and industrial engineers doing the estimating for quite simple products. Generally, however, the more complex the product, the greater the requirement for a technical background. The simpler the product situation, the more likely that Estimating is located under Cost Accounting.

Whatever their main background area, when estimators enter Estimating, there is much to be learned. Engineers have to become more familiar with the realities of production and the practices of accounting. Production people have to learn about cost accounting and usually more than previously necessary about engineering. Accountants have to learn more in greater detail about production and engineering.

I have always been less concerned about the experience of prospective estimators than about their personal characteristics and abilities. Usually, no matter what the background, any estimator with a good mind, an active curiosity, and enough energy can obtain a good working knowledge of the other fields along with experience on the actual estimating job. This is particularly true if management is willing to provide the necessary opportunity and training. The real danger is that such talent will be too burdened from the start with actual estimating workload to have time enough to develop knowledge in the other areas.

One other aspect of Estimating personnel bears mention. It unfortunately occurs too often. A company selects from within a person to be brought into Estimating. Assume in this case that he is a man. He is given extensive on-the-job training. His new boss and particularly his new coworkers serve as his mentors. They give him a lot of time. He is also

given time to be exposed to other management functions whose workings he previously had no opportunity to know. The company is spending a lot of money on his education. Perhaps he takes some courses or
even a degree at a local college. He does well, and more and more
independence is given him in developing estimates. Eventually he is a
qualified and good product-cost estimator. Being bright and curious, he
is not isolated in the company or the area. He meets other estimators
from other companies. If his salary scale is not kept commensurate with
his growing abilities, he is going to ask for more money. He recognizes
his increased value. If the company's pay scales and fringes in Estimating are not competitive, he may well leave. Most unfortunate is the
ensuing event in which the company has to pay the replacement the
salary requested by the estimator who left. Isn't the trained estimator
worth what we would have to pay his replacement? This happens not
only in Estimating but in other areas, e.g., in Production Planning and
Control, in Purchasing, and in other areas to which transfers from
within are frequently made. A company has a big investment, often
unmeasured and unrecognized, in good estimators. It pays to pay them
well. It is a point of which good management is aware.

Types of Estimators

Abroad, particularly in Europe   Abroad, particularly in Europe and
with product lines of some complexity, two types of estimators are
spoken of, budgetary estimators and detail estimators. Budgetary estimators estimate the cost of new products or new, advanced models of
the product line on which the final specifications, bills of material,
parts prints, etc., cannot be determined yet. This is conceptual or
parametric estimating. These estimators make the broad preliminary
estimate on which senior management and Marketing will determine
whether the new product or model will be viable in the light of possible
market pricing. On the other hand, detailed estimators do their estimating under conditions of specific product definitions. They work with
stated specifications, complete bills of material, and detailed parts
prints and make the detailed calculations of the costs of each operation
and each part. They are usually considered to need less technical education and detailed product knowledge than budgetary estimators,
though with experience and further knowledge and/or education, they
may work up to that position.

Such a distinction does not apply in this country. A large Estimating
department may have senior estimators who have junior estimators
under them, but the distinction is usually on the basis of knowledge

and experience and the assignment is usually in terms of product line, the more experienced and knowing estimator handling the more complex lines. Normally our estimators are held responsible for both types or areas of estimating. Thus a given product estimator will make budgetary estimates when and as needed but will also do the detailed estimate as the specifics of the product are developed.

Chapter

3

Cost Analysis and Identification

If you can express that of which you speak, and can express it in a number, you know something of your subject. But if you cannot measure it, your knowledge is meagre and unsatisfactory.

William Thomson (Lord Kelvin)

Since product-cost estimators deal with costs, they must know what kinds of costs occur in manufacturing and how these costs are identified objectively. Such identification demands good, i.e., objective, cost analysis. This chapter deals with the analysis and identification of costs and the single technique available for doing this important job factually instead of by guessing. All estimators should be knowledgeable in this area because they use cost data provided by other departments, particularly Accounting. In some estimating situations, estimators develop their estimates on the same basis and with the same data that Accounting used to analyze and identify the cost. Estimators should therefore be capable of evaluating the worth of the data provided, how well they were developed, and how realistically they were identified. Knowing how all this should be done properly makes estimators a constructive force for better cost data and better cost estimates.

Defining Cost Analysis

The term *cost analysis* is used in management in many different ways. For example,

• A corporation's purchasing executive or a government procurement analyst may study detailed quotations supplied by a number of vendors. The bids are broken down into details of material, labor, overhead, and profit. Each segment of cost is compared between the various bids and/or against previous actual costs on similar or identical work in an attempt to find out-of-line segments of costs and achieve a reduction in the cost being quoted.

• An industrial engineer studies the actual cost record of a given job or a given product to see what elements of cost are above standard or above estimate, to determine the most potentially profitable area for cost-reduction action.

Since analysis is the separation of a whole into its parts for study and interpretation, these two examples obviously can be called cost analysis. Historically, however, and in this book another meaning of cost analysis is studying the nature of an individual cost account in order to identify it objectively and thus be able to predict its response to changing operating conditions. Since estimators are in the business of predicting costs, cost identification is clearly of importance to them. I recognize that product-cost estimators normally do not do such cost-analysis work, which usually is the function of other management areas such as Cost Accounting and frequently Budgeting. However, since Estimating depends on Accounting for many of the data it uses in developing estimates, it is important that estimators be able to evaluate how cost data were identified and developed. Good estimators dare not know so little about the nature of operating costs that they must rely blindly on cost data supplied by another management area.

Need for Cost Analysis

As members of management, estimators should be aware of the many other uses and needs for this kind of cost analysis and identification. First, however, it is important to state the basic reason for the cost analysis we shall be doing, namely, to *separate the fixed from the variable costs.*

A company's management with a realistic knowledge of its fixed and variable costs by account, by product, and in total for a given sales forecast or for an actual sales mix can make better decisions. Some examples are given in Table 3-1, which is only a partial statement of management's need for good cost analysis and identification. The subject will have more and more direct bearing on the job of estimating in the years ahead, as discussed in Chapter 4.

Requirements for Good Cost Analysis

Proper cost analysis needs reasonably good data to analyze, which in turn requires well-defined and disciplined cost-charging practices and reasonably good cost reporting. Here are two examples of these requirements.

• Cost centers or departments must be fairly and accurately charged with the costs incurred in their operation. Thus, for example, electric-power costs must be charged to a given department on the basis of use. In many companies, it is reasonable to allocate total plant electricity costs to specific departments on some general basis like total horsepower of equipment in each department. But if one department does electric melting or heat treating, the cost of electricity has to be charged more carefully. Another example is the proper charging for supplies and/or perishable tooling. A requisitioning and charging system is needed if the using department is to be properly charged. Proper collection of actual costs provides good cost data, and that allows good cost analysis, identification, and estimating.

• Actual labor time reporting and piece counts must be correct. Although we may never reach absolute accuracy, we must achieve reasonable accuracy. Failure to do so can have a drastic effect on estimates. Many estimators use historical cost data, particularly operation costs, in developing costs for new models and new products. If cost data for specific production operations were calculated with fictitious labor times and/or fictitious piece counts, the estimate made with those data cannot be accurate.

Consider, for example, a problem that has been with us for years and continues to cause trouble in many companies, namely, bad piece counts. A production order for 150 pieces of a given part or subassembly is in the shop. There are 10 operations on the route sheet, numbered 10, 20, 30, etc. The shop reports that operation 10 completed 150 pieces; a little later operation 20 is reported complete for 149 pieces. Well, they scrapped a unit. Scrap occurs. Then operation 30 is reported finished for 275 pieces. We have the miracle of the loaves and fishes! When I mention that example at seminars, the heads nod and rueful smiles of recognition appear.

TABLE 3-1
Decisions Improved by Better Information about Fixed and Variable Costs

Marketing	Production	Top management (chief executive officer)
Determine contribution margins by product line and thus know better how to expend limited resources of marketing time and effort, i.e., which products to push	Calculate the projected effects of capital investments and of operational methods and changes	Make better long-range decisions on product contraction and expansion
If competition requires, decide which products can be marginally priced and the dollar range in which marginal pricing can be done	Decide make-or-buy issues on the basis of realistic calculations	Have operating budgets usable for *both* profit planning and cost control
		Calculate realistically the costs and effects of vertical and/or horizontal integration moves or mergers or acquisitions

If the individual production departments are on-line to a computer, so that operators punch their completed piece count, part number, and operation number into a direct entry unit, there is a solution. The computer can be programmed to compare this latest piece count with the piece count of the previous operation, and if it is not within a given limit of the previous operation's count, the computer will reject this fictitious entry. Most plants are not yet this far along with the computer. If the plant has departmental timekeepers or dispatchers who record completed operation counts on the route sheet copy accompanying the material, they can be made responsible for checking counts. But many plants with timekeepers or dispatchers do not have this checking routine built into their recording and reporting procedures, and many plants do not even use departmental timekeepers or dispatchers.

Most plants today do not have computerized piece-count checks. There is only one way I know to correct bad piece counts without them. When operation 30 is reported for 275 pieces completed, someone, usually from Production Planning and Control or from Cost Accounting, must immediately go to that department and ask the supervisor how such miracles can occur and what the correct piece count really is. Visits have to be promptly and emphatically made until supervisors are so tired of them that they will see to it that their operators report correct piece counts. It is a time-consuming job, but it has to be done. Fortunately, once you correct bad piece counts, they seem to stay reasonably accurate, and occasional lapses are easily corrected.

• Cost charging to specific cost accounts must be defined and done consistently. Labor, material, supplies, tooling costs, etc., must be charged to jobs and/or departments in a defined and consistent manner. Overhead allocations and fixed-cost charges, once logically settled upon, should not be arbitrarily changed.

• Each cost account must be considered and analyzed separately and individually. Amounts in more than one cost account cannot be lumped together and analyzed as a group. Only after they have been analyzed separately and identified objectively can similar accounts be grouped for use, if such grouping facilitates product costing and estimating.

Criteria for Identifying Costs

When you identify something, including costs, you are establishing categories. To categorize you need a basis. Thus, we must select a criterion or basis upon which to identify the costs of the enterprise.

One classification for grouping costs advocated in accounting and in management texts is *controllable* and *noncontrollable,* the implication being that we must accept some costs at their present level and cannot do anything about changing them or controlling them. This kind of thinking should be unacceptable to any manager worthy of the name. Managers must have the attitude that all costs are controllable. It just depends upon what level of management in what time frame can and will control the cost. This grouping and nomenclature are psychologically handicapping and of no real practical use.

Instead the useful way is to think of costs in terms of how they react to changes in activity. Thus, in a manufacturing enterprise, we would categorize costs according to how they do or do not change as sales and production activity change. This basis of cost identification is useful not only in manufacturing situations but also in service enterprises (banks, hospitals, museums, universities, etc.).

According to how they react to activity changes, there are three major types of costs: *fixed, pure variable,* and *mixed,* a mixed cost having both fixed and variable increments. Some accounting texts list four types of costs, based upon how they change with activity: fixed and semifixed, variable and semivariable. Since I have never understood the difference between semifixed and semivariable, I lump these two types under mixed costs.

Quantifying the Activity Measure

To be objective in cost analysis and identification, we must quantify not only our costs but also the activity measures used, which in turn depend upon the given situation and the given cost being analyzed. The activity measures are there. We may not have identified them and collected weekly or monthly data on them, but invariably usable activity measures are available for use.

Probably the most widely used manufacturing activity measure is *standard hours produced* if the company has labor standards for work measurement or even only for standard costing. For example, in the machine-tool and metalworking industries this is by far the most widely used activity measure in the production departments. It is also coming to be used more and more in the electronics industry. In industries with a lot of automatic machinery and one operator tending several machines, standard machine hours produced may be the applicable activity measure. In other manufacturing situations the usable activity measures may be in terms of products produced (gross, tons, linear feet, square feet, cubic feet, pounds, etc.). Activity measures are invariably available. They just have to be identified and counted. And this applies to service industries as well.

Certain requirements must be met by a usable and practical activity measure:

• It should be relatively unaffected by variable factors other than volume. For example, in metalworking *pieces produced* is an activity measure affected by the complexity of the pieces made. You can be twice as busy producing 200 complicated multioperation pieces as you are producing 2000 simple pieces. This is why most metalworking enterprises use standard hours produced and measure shop activity in those terms.

• Where possible, the best activity measure is one that can serve for many uses, particularly for both product costing and for budgeting. In many industries this is exactly the case for standard hours produced, but in other industries it is not possible. In the glass-bottle industry, for example, three or four activity measures are needed for product costing and budgeting: *hundredweight of material, tonnage poured,* and *standard hours produced.*

• The ideal activity measure also works for other costs in other departments and simplifies application in product costing. It is for this reason that standard hours produced is so widely used in metalworking industries. It generally is the one activity measure used in all production departments.

• Usable activity measures should be available from current records already being maintained so that no additional clerical or computer recording is needed.

Fixed Costs

This major category of costs is also called burden, overhead, constant costs, period costs, etc. A *fixed* cost is one that does not vary with changes in activity. Really there is no such cost. Any cost which we all would agree is fixed is actually fixed only for a given range of activity. Above that range, the cost will increase; and ideally if we fall below that activity range, we would manage the cost and see that it is reduced. For example, the well-known fixed cost of depreciation is fixed only for a given range of activity. If we boost production 30 or 40 percent, we usually need additional tooling, equipment, and possibly more workspace. Then depreciation costs would increase. Over a broad range of activity fixed costs will follow a broad step or staircase pattern.

Many managements have two difficulties with these fixed costs. The first is allowing the idea of "fixed" to imply "noncontrollable." It is only human, when identifying a cost as fixed, to consider it as unchangeable and beyond our control. Managers can never accept any cost as uncontrollable. A cost may be fixed and needed this year, but next year it may be avoidable or at least reducible. One of the great benefits of good cost analysis and identification for many companies has been the isolation of these fixed costs. Once identified and totaled as a group, their sheer money amount prompts management to give them a more intense and continued scrutiny than before.

The second, and even more serious, problem many companies and managements have with fixed costs is their incorrect allocation to different product lines. An example was given in Chapter 1 of a company with four product lines that allows one to be carried by the other three because of the way fixed costs have been allocated. Aside from the fact that this is all too common a situation, it is peculiar how the condition is often well recognized within the company without anything being done to correct the situation.

The work and money needed for better fixed-cost allocations is always productive, in my experience. For example, there are always valid marketing reasons for carrying product D ("we couldn't sell A, B, and C if we didn't have D"; "we need D to be competitive"; "it completes our product line"), but it is stunning how these reasons for carrying D diminish sharply in importance and frequency of mention once the real costs of product D have been determined. The problem is to persuade the management to find that out. When it is done, they make different decisions that are a lot more profitable. I am not necessarily advocating that product D be dropped. Often Marketing is entirely correct; they do need D to sell A, B, and C. I just urge that everyone know what D really costs. Then the management can and will make better decisions.

One unhappy manager gave me a gem of an example of improper allocation of fixed costs. "We allocate our Shipping department costs on the basis of sales volume. A major portion of our Shipping department cost is crating—the wood, the nails, the labor. Most of our sales volume is in a product line that doesn't get crated!"

Pure-Variable Costs

The second major type of cost is the *pure variable*. To be a pure variable (1) the cost must be incurred only when activity starts; thus, the cost begins only when activity begins, not when the organization or plant is established. (2) There must be a direct relationship between the cost and the activity. As activity increases, the cost increases; as activity drops the cost drops.

In a manufacturing company we deal mostly with production and sales activity. For a service company, it is operational and revenue activity. Examples of obvious pure-variable costs in manufacturing companies are commission costs, which vary with sales activity, and direct raw-material costs, which vary with production activity.

Usually these pure-variable costs have not only a *direct* relationship with activity but also a *direct, incremental* relationship with activity. Consider raw-material costs, for example. They vary both directly and incrementally with production. Thus, if there is a dollar's worth of material in every unit made, the cost will rise a dollar for each additional unit produced. In contrast, although commission cost is a pure-variable cost, it frequently does not have a direct incremental relationship to sales activity and sales volume because many companies have a declining-commission schedule as order size increases.

Notice how these fixed and pure-variable costs switch roles when we consider them on a per product unit basis instead of as discrete cost accounts. The fixed cost becomes variable and the variable cost becomes fixed. Thus, the pure-variable cost of direct material, say $1 per unit, is fixed per unit. But the fixed cost, on a per product unit basis, declines as production increases and increases as production falls. Depreciation, for example, is a well-known fixed cost. But if the depreciation on a given machine is $10,000 per year and you produce 10,000 units on that machine in the year, the proper depreciation cost per unit is $1.00. If you produce 100,000 units, the cost per unit is $0.10. If you produce 5000 units, the cost per unit is $2.00. This all seems obvious, but it is surprising how managers can lose awareness of this fact and fall into illogical cost decisions as a result.

An important specific cost in manufacturing is direct labor. Unlike direct raw material, this cost is not always a pure variable. In some production situations it is just not practical to consider and treat it as a pure variable. In chemical plants, oil refineries, and heat-treating departments, for example, production can rise or fall 20 to 30 percent and you do not have to add labor, nor can you reduce the work force. But in the vast majority of production situations, direct labor should be treated and accounted for as a pure-variable cost. For example, in most metal-working plants today, a turret-lathe operator who has no work for the day will not be sent home. Such workers are too hard to find. Some indirect labor task will be found to utilize the operator's time and day's pay. A good supervisor will have such work set aside for just such a contingency. But if the operator is not charged to indirect labor, two cost accounts will be distorted. The indirect work done in the day will not be charged, and the direct labor not done will be overcharged. Thus properly accounted for, direct labor, in most manufacturing situations, can and should be a pure-variable cost. Proper labor-cost reporting is also good operating procedure because, in practice, variable costs tend to receive more management attention and stronger management control.

Of the three major types of costs, pure-variable costs are commonly the highest percentage of total costs in terms of money but the lowest in terms of cost accounts. Many cost accounts that appear to be pure variable are, in fact, mixed costs.

Mixed Costs

The third and final major category of costs is *mixed costs*, sometimes called *semivariable costs*. They are the ones that have a fixed increment and, in addition, a variable increment that increases or decreases directly with activity. Most of the errors made in cost identification are made in distinguishing between the pure variable and the mixed costs. Many costs that are in fact mixed are misidentified as pure variables.

Apparently the most popular technique for identifying the mixed costs versus pure variables is guessing. In this day and age we should try to reduce the management areas in which we guess and should quantify where we can. Many areas in management call for experienced judgment since we lack or cannot afford to collect the facts and quantifications we really need. But modern management should not guess as much as it does in this all-important area of product costing. In this section, we consider the quantifiable way to identify the three subtypes of mixed costs.

MIXED LINEAR COSTS

A *mixed linear cost* is a cost that has a fixed increment and on top of that a variable increment that increases the same amount with each equal increase in activity. For example, consider the cost of operating supplies in a metalworking plant. Operating Supplies is the cost account into which, normally, Cost Accounting charges the miscellaneous supplies required to run the manufacturing operation (coolant fluids, floor-cleaning compound, safety goggles, etc.). Every enterprise, certainly every manufacturing enterprise, has such an account.

What kind of cost is it? First let's do a little guessing. If the plant is not producing, we should have no cost for operating supplies. When it is producing, the greater the production or output, the greater the cost of operating supplies should be. We are guessing that this is a pure-variable cost, which begins only when production (activity) begins and increases directly with production or activity. The odds are overwhelming that we are wrong. In 35 years of experience, in industry after industry, I have never seen this particular cost act as a pure variable as it apparently should. It has always been a mixed linear cost.

Instead of guessing, we should analyze such a cost quantitatively and identify it factually. A suitable technique is available to do just this. Consider the data in Table 3-2. The activity measure is standard hours, and we have the cost of operating supplies for 12 successive months. Clearly, we have here two variables, the independent variable of activity, as measured in terms of standard hours produced, and the presumably dependent variable of dollars of cost of operating supplies. We can picture the relationship between these two variables by the simple procedure of a *scatter diagram*.

On a piece of arithmetic graph paper, the longer (horizontal) axis is used as the x axis and the vertical (shorter) axis, as the y axis. We scale from zero along the x axis the activity measure, in this case standard hours produced. We scale from zero up the y axis the dollars of cost. Then we plot the 12 months of actual cost versus activity (Figure 3-1).

The plot shows what we would expect. There is a relationship, or correlation, between activity and dollars of operating supply cost. If the correlation were perfect, the plotted points would fall on an ascending straight line, but this happy situation is never seen in real life. There is always a scatter of the plotted points. What line through these plotted points best describes the relationship between these two variables and best identifies what kind of cost this is?

TABLE 3-2
Actual Cost of Operating Supplies at Actual Operational Activity

Month	Standard hours produced (000)	Cost of operating supplies
January	29	$4400
February	30	3600
March	31	4100
April	37	5200
May	35	4600
June	40	5500
July	20	3400
August	23	4200
September	32	4600
October	33	5000
November	24	3600
December	25	3800

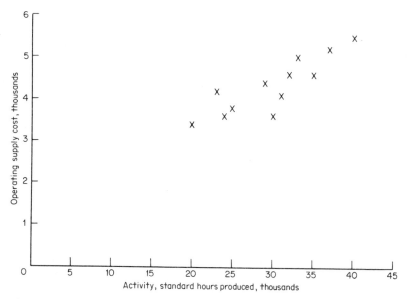

Figure 3-1 Operating supply cost versus standard hours produced.

One approach is to draw, by eye, a line through the plotted points so that there are approximately as many points above the line as there are below the line. This is called *eyeballing the line*.

The eyeballed line drawn in Figure 3-2 seems reasonable and representational of the relationship between these two variables. (Actually this is not the best way to draw the line. Later in the chapter we shall examine a better way, but for the moment let's accept this line as usable and descriptive of the relationship.) Obviously this is not a pure-variable cost. A variable cost would start at zero dollars for zero activity and rise in dollar cost as activity increases. If we drew such a line on the graph, it would not appear to most eyes nearly as descriptive as the line in Figure 3-2. Accepting the eyballed line for the moment, we see that this is not a pure-variable cost as we guessed. It is, in fact, the first type of mixed cost, a mixed linear cost. It has a fixed increment and above it a variable increment that increases with activity.

The formula for such a line is

$$y = a + bx$$

where y = dollars of cost at given activity
$\quad\quad a$ = fixed increment
$\quad\quad b$ = slope of line
$\quad\quad x$ = any given activity

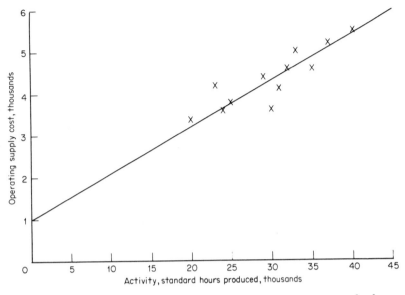

Figure 3-2 Eyeballed line for operating supply cost versus standard hours.

With the eyeballed line used in Figure 3-2 we can readily calculate the value of a and b, from which we can calculate the y value, or dollars of cost, for any x value, i.e., activity level, we want to use. The a value we can read off the graph. It is the dollars of cost where the line crosses the y axis at zero activity or zero x. In Figure 3-2 the a value is $1000. The b value, or slope of the line, can be calculated by the method of differences for determining the slope of a line. At zero activity, the line is at $1000. At 45,000 standard hours produced the line is at $6000 of cost. Thus:

$$\begin{array}{rl} \$6000 \text{ at} & 45{,}000 \text{ standard hours produced} \\ \underline{-1000} & \underline{\quad -0\quad} \\ \$5000 \text{ at} & 45{,}000 \text{ standard hours produced} \end{array}$$

Thus the line went up $111.11 ($5000/45) for each additional thousand hours of activity. And the formula in our terms is

$$y = a + bx$$

Operating supply cost
$$= \$1000 + (\$111.11/1000 \text{ standard hours produced})$$

The trouble with eyeballed lines is that different analysts can visualize and draw different lines to describe a given relationship. It is much better to draw the line by least-squares line fitting, which ensures that everyone will arrive at the same line, no matter who calculates and draws it. In serious cost analysis on large costs, it is best to avoid fitting a line by eye and to use least squares. The procedure for least-squares line fitting is given later in this chapter.

In discussing pure-variable costs versus mixed linear costs one subtlety is worth reviewing because of a difference between what should be and what actually is. Consider the example of operating supplies. There is no explainable reason why this specific cost should be other than a pure variable. Why should we have $1000 of fixed cost per month when the plant is not producing? Faced with this identification of operating supplies as a mixed linear cost, the theorist will argue that it should not occur, that the cost should be treated as a pure variable, and that, at best, the line describing the cost should be drawn from zero up through the points. The result of such a line versus the mathematically fitted line (and even the eyeballed line) will be to understate the cost at lower levels of activity and to overstate the cost at high levels of activity.

This is illustrated in Figure 3-3, which shows that if we identify and treat this mixed linear cost of operating supplies as a pure variable, we shall predict too low a cost at the lower levels of activity and too high a cost at higher levels of activity. The busier we become and the greater

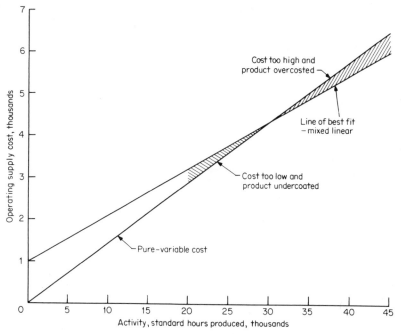

Figure 3-3 Results of confusing mixed linear and pure-variable costs.

the activity level, the greater the error in our overprediction of the cost. This is a simple illustration in terms of only one cost, but when many mixed linear costs are erroneously treated as pure variables, the product-cost effects and errors may become substantial. When the confusion occurs in budgeting, the budget allowances can be seriously unjust, too tight at lower levels of activity and too loose at higher levels of activity.

This conflict between what should be and what actually happens should not cause us concern. What we are really interested in is how a cost reacts in real life, within the range of activity with which we actually have to deal. If our evidence indicates that a given cost is a mixed linear cost, we should so identify it for use in product costing and estimating.

MIXED STEP COSTS

A *mixed step cost* is a cost that increases in a step pattern as activity increases. The cost is fixed for a certain range of activity, but when activity increases above that range, the cost increases and is fixed for a limited further range of activity, and so on. The cost increases and the range of activity for which the cost remains fixed may or may not be constant at successive steps. This type of cost is characteristic of indirect labor costs in an enterprise with good indirect-labor cost control.

For example, in a metalworking shop the activity measure is standard hours produced and the cost in question is setup labor. In this shop it is an indirect labor account, and the cost record for the most recent 12 months is shown in Table 3-3. Graphing these dollars of cost against activity on arithmetic graph paper gives Figure 3-4.

TABLE 3-3
Actual Cost of Setup Labor at Actual Operational Activity

Month	Standard hours produced (000)	Cost of setup labor
January	29	$4800
February	30	5000
March	31	5100
April	37	5800
May	35	4900
June	40	6100
July	20	3800
August	23	4300
September	32	4800
October	33	5200
November	24	3900
December	25	4000

The plotted points reveal a useful correlation between dollars of setup-labor cost and activity, indicating good cost control in this area of labor cost. When you plot dollars of a given indirect labor cost against activity, you often find a relatively flat line with perhaps higher costs at the busy periods to reflect extra overtime in those busy months. Such a flat line raises the question: Why do we need the same indirect labor costs in the slower months as in the busy months?

From the plot in Figure 3-4 it is apparent that a straight line could be drawn through the points and this cost could be identified as a mixed linear cost. If a mechanical approach to identifying this cost is taken, that might well be done. And it would be wrong. We are dealing with cost analysis, and to do meaningful analytical work, we cannot be mechanical. We must bring our knowledge of the operation to bear. In slow months, we have the fewest setup workers. As we get into our busier months, we must take the better operators and use them as setup workers or hire setup workers. This cost is best treated as a mixed step cost, as illustrated in Figure 3-5. Thus, with this particular mixed step cost, the cost of setup labor would vary as shown in Table 3-4.

Where one step ends and the other begins and the level of each step are all apparent in this example, but this is not always the case. Cost analysts must often use their judgment in deciding where to cut one

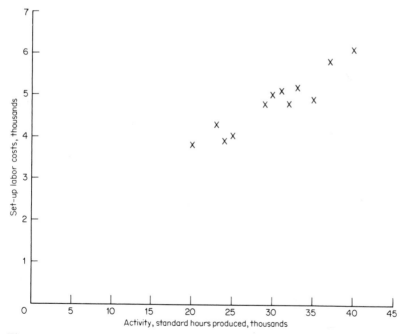

Figure 3-4 Setup labor cost versus standard hours produced.

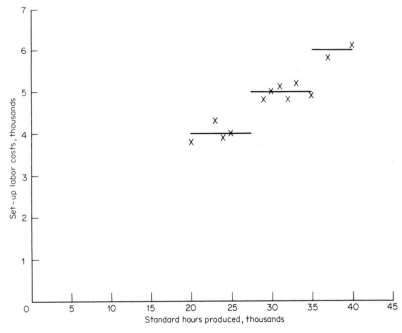

Figure 3-5 Mixed step cost, setup labor versus standard hours produced.

step off and begin the next and at what level the next step is located. This can be particularly important in budgeting situations, where budget allowances will be determined on the basis of the judgments made. Equally pertinent in product costing is that such judgments will affect the data made available to Estimating. Although we cannot eliminate human judgment in the analysis of this kind of cost, we can

TABLE 3-4

Standards of Setup Cost at Various Levels of Operational Activity

Production, standard hours	Setup labor cost
20,000–27,499	$4000
27,500–34,999	5000
35,000–40,000	6000

and should apply our judgment on the facts we have available instead of simply guessing what type of cost it may be.

MIXED CURVILINEAR COSTS

The third and last type of mixed cost, *mixed curvilinear cost,* increases with activity, above a fixed increment, at an increasing or decreasing amount per unit of activity measure. If the cost increases at an increasing amount, it plots as in Figure 3-6*a*. If the cost increases at a decreasing amount, it plots as in Figure 3-6*b*. Descriptive lines can be mathematically fitted to such curvilinear relationships. The specific formula depends upon the actual relationship between the two variables of activity and dollars of cost.

In real-life cost analysis and product costing, this type of cost is so rare that in 35 years of practice I have never used one. The reason for their rarity is that seldom do we have such accurate cost reporting that we can fit curvilinear relationships to dollars of cost versus units of activity. Many of us have fitted straight lines to data that we suspect should be curvilinear, but as we improve our cost reporting and make more use of calculators and computer terminals, we are more likely to find curvilinear cost-activity relationships.

JUSTIFICATION OF MIXED COSTS

Some writers advise against using mixed costs, arguing that it is entirely practical to treat any given cost account as either fixed or as pure variable. They feel that mixed costs are difficult to explain and tend to confuse managers, particularly at the lower levels of management. They admit that mixed costs exist but want to see them categorized as fixed or variable costs depending on the main portion of their dollar amount. In my judgment this is too coarse an approach. I have seen many mixed costs amounting to so much money in the aggregate that unnecessary costing and budgeting errors result if they are not handled as mixed costs. I have had no trouble over the years in explaining the graphing technique to a production supervisor. In fact, in budgeting situations, I have found it to be a clear and effective way of convincing supervisors that the budget allowances so determined made sense and were fair.

Using the Past

The chief criticism of the technique advocated above for the analysis and identification of costs is that it uses the past as a predictive base for the future and is in danger of perpetuating past inefficiencies. Are we justified in using past actual performance in estimating and cost analysis?

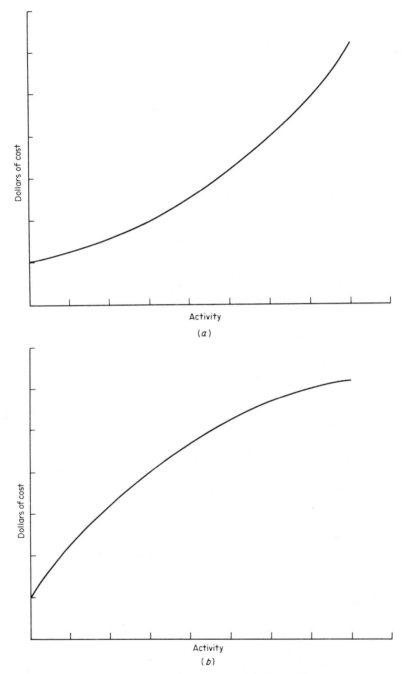

Figure 3-6 (*a*) Concave and (*b*) convex mixed curvilinear cost.

As a student and practitioner in both the estimating and budgeting fields, I have often run into this criticism. What constructive suggestion is offered in place of this destructive criticism? Invariably it is subjective human judgment, which I call guessing. I wish I could present six different alternatives. They do not exist. At the present state of the art of cost analysis we either guess or do what is proposed above, and I want to reduce the guesswork. There is so much roughness, so many suppositions made in costing work, that we should quantify wherever we can.

In costing, whether for estimating or for cost analysis, the real problem often is not having decent records available. When we have such past-performance data that we consider usable, we should use them. When we don't have them, Estimating should be among the leaders urging management to see that these data are collected properly and promptly.

Of course, the estimators and the cost analysts can never use past history blindly. They must be able to evaluate its accuracy, be familiar with how it was collected, and know what such data do and do not include and the surrounding operating conditions. Only then can they deem it usable or unusable for present and future operating conditions.

Estimating's Use of Cost Analysis Facts

Although product-cost estimators normally do not analyze and identify costs, in many industries and many estimating situations the quantifications of cost analysis work can find direct application in product-cost estimating. The activity measures used to analyze the costs, for example, are the same measures on which those costs are applied and built up to develop the product-cost estimate.

Before looking at some specific examples let's consider the most common estimating situation. An estimator develops a prediction of how many labor hours or machine hours will be needed to produce the parts required in the final product being estimated and then applies a plant burden rate or, better, departmental burden rates. Or Accounting applies them on the estimator's developed time values. In either case, usually Accounting develops the burden rates to be applied. These rates are needed to cover the cost of such things as plant building costs, production management salaries, machinery depreciation, production staff services, such as production planning and control, industrial engineering, power, operating supplies, indirect labor, etc. The burden rates to be applied are the quotient found by dividing the budgeted or past actual costs for all these overhead cost accounts by the labor or machine hours needed to make the sales forecast.

Assume that this cost analysis has been correctly done. We now have activity measures and costs related to them. Even better, we can segregate such costs and the totals of many such cost accounts into their variable increments and into their totals of fixed costs. The fixed costs and the fixed increments of mixed costs we can apply as burden rates, as most of us do presently for *all* overhead costs. But now, as a result of our cost analysis, we can apply the variable increments to the product cost on the basis of the activity measure or measures. In terms of the operating supplies cost used as an example of a mixed linear cost earlier in this chapter, if we estimate a job to take 150 standard hours of work, we can estimate operating supplies cost on the job at 150($0.111) = $16.67. In actual application, we would be using the total variable increment of many cost accounts that were measured against the given activity measure. That is, we are applying the variable portion of our manufacturing overhead in the same manner as we apply such pure variables as direct material or direct labor.

In many industries, companies that have properly analyzed and identified their costs do not include the variable segments of their mixed costs in one overall burden rate for the plant or in specific burden rates for each specific cost center. Instead they apply the variable increments of overhead to product costs on the basis of the activity measures, which are used not only to analyze and identify costs but also to build up the product-cost estimate. For example,

• In some metalworking plants, variable manufacturing overhead increments are applied against standard labor hours estimated for the parts needed in the final product.

• In some electronics companies, estimated printed-circuit-board assembly costs and wire-harness assembly costs are estimated in terms of labor hours. Then variable manufacturing overhead costs are applied on the basis of these standard labor hours.

• In some wire-drawing plants, variable overheads are applied in Estimating to standard machine hours produced, which is the activity measure used to analyze and identify the costs.

• In some corrugated-box plants, variable manufacturing overhead is applied by Estimating to machine hours needed to produce the order.

• In some glass-bottle plants, different segments of the variable cost increments of manufacturing overhead are applied on the basis of three activity measures: hundredweight of material used, tonnage poured, and standard machine hours produced.

In all these examples, the activity measures used to analyze and identify the various costs were the same ones used to apply the variable increments of overhead in developing the product cost estimate.

Simple Statistical Tools for Cost Analysis

Today's tremendous variety of computation tools presents a danger unless we understand what the machine is doing. In this section, we review certain statistical tools pertinent to cost analysis work and manually make the calculations necessary for their development and use. Once we understand these calculations, we can safely let the calculator or computer do the work.

All readers are urged not to skip this section. A little refresher never hurt anyone. While the examples lie in the cost analysis area, the tools themselves are applicable in many estimating situations (see Chapter 7).

The specific statistical tools are least-squares line fitting and the coefficient of correlation, both tools of regression or correlation analysis.

LEAST-SQUARES LINE FITTING

In our discussion of mixed linear costs we were faced with the problem of describing the relationship between two variables, the production activity measure of standard hours produced, and the cost of operating supplies. The relationship was shown in Figure 3-1. In that section, we eyeballed a line to fit the scattered points but said that it was far better to use the line of least squares. This *line of mathematical best fit* is the line for which the sum of the squares of the distances of the plotted points about the line is a minimum.

COEFFICIENT OF CORRELATION

To the experienced cost analyst or estimator it is obvious that there should be a relationship or correlation between activity and the cost of operating supplies, and in the example shown in Figure 3-1 there certainly is. As activity increased, so did the cost of operating supplies. But we often need to quantify that relationship. Is the relationship close enough to be usable for identification and predictive purposes? In other circumstances, other questions may arise. For example, in a cost analysis and identification situation, we may have several possible activity measures to use against dollars of cost. Which activity measure is best?

Help in answering such questions can be given by the *coefficient of correlation r*, which gives a numerical value to the degree of relationship between two variables. If the correlation is perfect, the coefficient of correlation is 1.0 and the plotted points will fall exactly along a straight line on the plot. Or if the best description of the relationship is curvilinear and the coefficient of correlation is 1.0, the plotted points will fall right along a curved pattern on the plot. We pointed out earlier that this never happens in real life. There is always some scatter, and the coefficient of correlation is less than 1.0. The coefficient of correlation is interpreted as follows:

1.00–0.80	High relationship
0.80–0.60	Marked degree of relationship
0.60–0.40	Moderate degree of relationship
0.40–0.20	Low degree of relationship
0.20–0	No, or negligible, relationship

Simple and Multiple Regression

We are dealing here with *simple regression*, the measure of the relationship between *two* variables. *Multiple regression* measures the relationship between more than two variables. This more complex correlation analysis is used in mathematical approaches to sales forecasting or in operations research work. It is rarely found in costing and estimating work at present, but Estimating will be using it more in the future (see the section Multiple Regression in Chapter 7).

Spurious Correlation

There is such a thing as *spurious correlation*, in which there is a high correlation coefficient (0.80 or greater) between two factors although there is no *causal* relationship. For example, at one time, there was over a 0.90 correlation coefficient between the per capita consumption of alcohol in the United States and the average salary of clergymen. In the area of sales forecasting using leading indicators, some spurious correlations have been found, but they are rarely a danger in normal cost analysis work.

Coefficient of Determination

A variant of the coefficient of correlation r is the *coefficient of determination* r^2, which is simply the coefficient of correlation squared. If the calculator is programmed to calculate r^2, using the square-root key will yield the coefficient of correlation. In this discussion, all calculations are for the coefficient of correlation.

EXAMPLE OF LEAST-SQUARES LINE FITTING AND CALCULATION OF THE COEFFICIENT OF CORRELATION

For the mixed linear cost of operating supplies, instead of eyeballing a line, let us calculate both the line of least squares and the coefficient of correlation. The activity and cost data in Table 3-2 are repeated for convenience as Table 3-5.

Since we scaled activity (standard hours produced) along the x axis because it is the independent variable, we can call the total of that variable Σx; it has the value 359 in this example. Because cost of operating supplies is the dependent variable, it is scaled along the y axis. Therefore the total cost is Σy, and in this example it is $52,000.

When we algebraically manipulate the basic straight-line formula $y = a + bx$, we get

$$b = \frac{N(\Sigma xy) - \Sigma x(\Sigma y)}{N(\Sigma x^2) - (\Sigma x)^2} \quad \text{and} \quad a = \frac{\Sigma y - b\Sigma x}{N}$$

where N is the number of occurrences. The formula for the coefficient of correlation is

$$r = \frac{N(\Sigma xy) - \Sigma x(\Sigma y)}{\sqrt{N(\Sigma x^2) - (\Sigma x)^2}\sqrt{N(\Sigma y^2) - (\Sigma y)^2}}$$

TABLE 3-5

Actual Cost of Operating Supplies at Actual Operational Activity

Month	Standard hours produced (000)	Cost operating supplies
January	29	$ 4,400
February	30	3,600
March	31	4,100
April	37	5,200
May	35	4,600
June	40	5,500
July	20	3,400
August	23	4,200
September	32	4,600
October	33	5,000
November	24	3,600
December	25	3,800
Total	359	$52,000

TABLE 3-6
Arithmetic Data for Least-Squares Line Fitting and Determination of Coefficient of Correlation for Operating Supply Cost

Month	Std. hours prod. x (000)	Cost y	xy	x^2	y^2 (000)
Jan.	29	$ 4,400	127,600	841	19,360
Feb.	30	3,600	108,000	900	12,960
Mar.	31	4,100	127,100	961	16,810
Apr.	37	5,200	192,400	1,369	27,040
May	35	4,600	161,000	1,225	21,160
June	40	5,500	220,000	1,600	30,250
July	20	3,400	68,000	400	11,560
Aug.	23	4,200	96,600	529	17,640
Sept.	32	4,600	147,200	1,024	21,160
Oct.	33	5,000	165,000	1,089	25,000
Nov.	24	3,600	86,400	576	12,960
Dec.	25	3,800	95,000	625	14,440
Total	359	$52,000	1,594,300	11,139	230,340

It may look formidable, but it is basically easy to use. We already have Σx and Σy, and we know that N, the number of occurrences, is 12 in this example. We must calculate Σxy, Σx^2, and Σy^2. The results are given in Table 3-6.

Now we can substitute figures for symbols in the formulas:

$$b = \frac{N(\Sigma xy) - \Sigma x(\Sigma y)}{N(\Sigma x^2) - (\Sigma x)^2} = \frac{12(1,594,300) - 359(52,000)}{12(11,139) - (359)^2}$$

$$= \frac{463,600}{4787} = 96.85$$

$$a = \frac{\Sigma y - b(\Sigma x)}{N} = \frac{52,000 - 96.85(359)}{12}$$

$$= \frac{17,232}{12} = 1436$$

Thus, the formula for this mixed linear cost is

Operating supplies cost
= $1436 + ($96.85/100 standard hours produced)

Contrast this with the eyeballed line and its formula

Operating supplies cost
= $1000 + ($111.11/1000 standard hours produced)

The line drawn by eye was guesswork; the least-squares line is quantified and calculated; you, I, and everyone else will arrive at the same answer.

To calculate the coefficient of correlation

$$r = \frac{N(\Sigma xy) - \Sigma x(\Sigma y)}{\sqrt{N(\Sigma x^2) - (\Sigma x)^2}\sqrt{N(\Sigma y^2) - (\Sigma y)^2}}$$

$$= \frac{12(1,594,300) - 359(52,000)}{\sqrt{12(11,139) - (359)^2}\sqrt{12(230,340,000) - (52,000)^2}}$$

$$= \frac{19,131,600 - 18,668,000}{\sqrt{4787}\sqrt{60,080,000}} = \frac{463,600}{536,286} = 0.86$$

An 0.86 coefficient of correlation indicates a high degree of relationship between the two variables. We have quantified our judgmental evaluation that the two variables are related.

High-Low Method of Line Fitting

A very poor method of describing the relationship between two variables occasionally used is called the *high-low method*. With this method you take the months of lowest and highest activity and their respective dollars of cost. Then for both activity and dollars of cost you subtract the lowest from the highest to calculate the amount of dollar increase per unit of activity increase. Using the amount of dollar increase, you calculate back from either the busiest or slowest month to determine the costs' fixed increment.

To do this in the case of the operating supplies cost (Table 3-2) the calculation would be

Standard hours produced	Cost
High: June 40,000	$5500
Low: July 20,000	3400
Difference 20,000	$2100

Thus

$$\frac{\$2100}{20} = \$105/1000 \text{ standard hours}$$

June:

$$\$5500 - 40(\$105) = \$5500 - \$4200$$
$$= \$1300 \text{ calculated fixed cost}$$

July:

$$\$3400 - 20(\$105) = \$3400 - \$2100$$
$$= \$1300 \text{ calculated fixed cost}$$

In this case the formula would be

Operating supply cost
$$= \$1300 + (\$105/1000 \text{ standard hours produced})$$

In contrast the formula developed by the least-squares method is

Operating supply cost
$$= \$1436 = (\$96.85/1000 \text{ standard hours produced})$$

In this example, the high-low method would yield too steep a line, which would be too low at low levels of activity and too high at higher levels of activity; the busier the shop, the greater the overestimation of this cost. In this example, the differences are not tremendous, but I have seen cases where this method resulted in serious differences from the least-squares line.

This high-low method's only advantage is its simplicity. Unfortunately, it is also simple-minded. It is completely inadequate for serious cost analysis work because it uses only 2 of the 12 months of data, thus ignoring 83 percent of the data. Least-squares analysis is too easy ever to justify using such an inadequate method as the high-low approach.

Case Problems in Cost Analysis and Least Squares

Nothing can replace hands-on work when dealing with a technique. If you do cost analysis and/or regression analysis as part of your daily job, these case problems can be skipped, but if it has been 6 months or more since you had to handle such matters, it can do no harm to take your hand calculator and try these two case problems. Management is becoming more and more involved with statistics and its techniques. This is an opportunity to get some hands-on involvement yourself. The answers follow at the end of the chapter.

PROBLEM 1

A large metalworking shop measures its operational activity in terms of standard hours produced. A substantial cost is Perishable tooling. The record of this cost and shop activity for 12 months is given in Table 3-7.

(a) Plot these data on arithmetic graph paper (preferably 10 to the inch), with activity (standard hours produced) on the x axis and dollars

TABLE 3-7
Actual Cost of Perishable Tools at Actual Operational Activity

Month	Standard hours produced (000)	Cost of perishable tools
January	38	$ 5,600
February	37	4,600
March	39	5,500
April	43	5,600
May	42	5,700
June	45	6,000
July	31	4,400
August	33	4,300
September	40	5,200
October	41	5,800
November	36	5,300
December	34	5,000
	459	$63,000

of perishable tooling cost on the y axis. Eyeball a line through the plotted points.

(b) Now calculate the least-squares line and plot it on the scattered diagram you made for **(a)** above.

(c) Calculate the coefficient of correlation.

PROBLEM 2

In Table 3-3 the cost data for setup labor for 12 successive months were given. We saw that in this and other indirect-labor areas the best single way to identify such a cost is as a mixed step cost, because it rises or falls in terms of employees, or by steps. To increase your familiarity with least squares and the coefficient of correlation, a line can be calculated for these cost-activity data.

(a) Calculate the least-squares line. What is the formula for this cost if it is treated linearly?

(b) Calculate the coefficient of correlation.

Answers to Case Problems

PROBLEM 1

(a) The plotted points of perishable tooling cost versus the activity of standard hours produced are shown in Figure 3-7. Notice how easy it

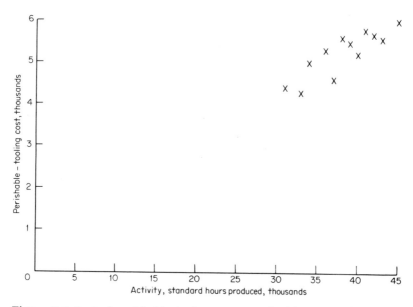

Figure 3-7 Cost of perishable tools versus standard hours produced.

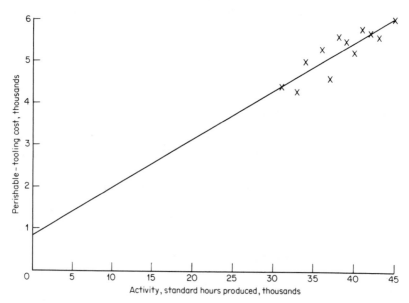

Figure 3-8 Plot and least-squares line for perishable tooling versus standard hours produced.

would be to eyeball a line from zero up through the 12 plotted points, in which case, we would be saying that it is a pure-variable cost.

(b) This cost is, in fact, a mixed linear cost. When you calculate the values needed for the least-squares formulas and for the coefficient of correlation you obtain

$$N = 12 \qquad \Sigma xy = 2,432,600$$

$$\Sigma x = 459 \qquad \Sigma x^2 = 17,755 \qquad (\Sigma x)^2 = 210,681$$

$$\Sigma y = 63,000 \qquad \Sigma y^2 = 334,240,000 \qquad (\Sigma y)^2 = 3,969,000,000$$

$$a = \$841 \qquad \text{and} \qquad b = \$115.26$$

Thus the least-squares formula for the line is

Cost of perishable tooling
$$= \$841 + (\$115.26/1000 \text{ standard hours produced})$$

The least-squares line is plotted in Figure 3-8.

(c) The coefficient of correlation is 0.87, which indicates a very usable relationship.

PROBLEM 2

(a) The values for the least-squares formula and for the coefficient of correlation for this setup labor cost are

$$N = 12 \qquad \Sigma xy = 1,771,100$$

$$\Sigma x = 359 \qquad \Sigma x^2 = 11,139 \qquad (\Sigma x)^2 = 128,881$$

$$\Sigma y = 57,700 \qquad \Sigma y^2 = 283,120,000 \qquad (\Sigma y)^2 = 3,329,290,000$$

$$a = \$1440 \qquad \text{and} \qquad b = \$112.58$$

The least-squares formula for the line is

Cost of setup labor
$$= \$1140 + (\$112.58/1000 \text{ standard hours produced})$$

(b) The coefficient of correlation is the very high one of 0.94.

The Influence of Accounting and Direct Costing

Chapter 4

Since Estimating depends on Accounting for a substantial part of the data it uses to develop product-cost estimates, it is advisable for estimators to know accounting. By this I do not mean T accounts and double-entry bookkeeping, asset, liability, capital, and cost accounts and how they are used to develop the balance sheet and the profit and loss statement. Everyone in management should have a grasp of these matters and can readily acquire it, if necessary, from books, night and weekend courses, and management seminars. Instead, estimators have to deal with cost accounting and overhead allocations, much more subtle matters, varying with the company and the product.

Every estimator should also be aware of the concept of direct costing, how it differs from the conventional absorption costing that most companies use today, and how it affects the structure of the estimate itself.

Estimating vis à vis Accounting

Earlier I said that every manufacturing company needs an Estimating department whose task it is to provide management with an *independent, realistic prediction* of what it will cost to make the product. This seemingly simple definition of an estimate has wide ramifications in actual application. In developing the cost of the product, costing and burden rates have to be applied to determine the total cost properly. Very rightly, it is Accounting that collects actual cost data and calculates costing rates in terms of money and the burden percentages that will be used. These inputs obviously have a great effect on the total product-cost prediction that is finally developed and used by management. For example, the estimator can do exhaustive research and painstaking calculations of the labor and/or machine hours required to make the product, but if the costing and burden rates are gross averages, improper, and/or inaccurate, prediction of the final calculated total product cost will be wrong. Furthermore, management will have wrong data on which to base important decisions.

Clearly, we often have a division of the real responsibility for the quality of the final total estimate. This division can be harmful to the company because management has no one department to hold responsible for the accuracy of the predicted total cost to make the product. I believe that the most practical resolution of the problem is to have Accounting and Estimating *jointly* responsible for the costing and burden rates used on the estimates. Thus, Accounting would still do all the actual cost collection, costing, and burden-rate calculations, but Estimating would be equally responsible with Accounting for ensuring that the rates are properly and logically calculated and are the ones to be applied properly in developing total product cost. Estimating would have to review, criticize, and perhaps participate in establishing the logic applied to achieve any necessary corrective action.

Such a proposal for such joint responsibility is anathema to many a controller, senior manager, or estimator. Many controllers would consider it a gross intrusion into what they see as Accounting matters. Many estimators would consider it an imposition to require their participation in such decision making in an area they do not view as their responsibility. But isn't a good organization a structure of checks and balances? Shouldn't different, independent thinking be brought to bear on these important costing and burden rates? If this is done, isn't the company more likely to have better costing and burden rates applied in the estimates that are developed? If it is done this way, senior management can justifiably look to Estimating to bear the responsibility for realistic product-cost estimates.

This discussion is not germane to your company if you have good estimating and logical product costing and senior management can, with justice, expect Estimating to be responsible for realistic estimates. But I have seen too many companies where:

Some products are carried by other products because the product costs are not realistic.

Overhead allocations are illogical and thus too high and unfair for some products and unrealistically low for others.

Costing rates are so grossly averaged that they bear no really usable relationship to actual departmental costs and how they should be applied to specific products.

Estimators *blindly* use costing and burden rates without any real understanding of how those rates were constructed and what they contain or any voice in how they should be calculated and applied.

Carefully constructed estimates of labor or machine hours have costing and burden rates applied to them which only Accounting understands and had a part in developing and which are not logical.

The list of observed situations could be extended. The whole point is that Estimating should have a voice in how the costing and burden rates used in the estimate are developed and applied. *Then* senior management can hold Estimating responsible for the quality of the total cost prediction. In turn, Estimating must understand the realities of product cost accounting. If the rates given Estimating are inadequate, Estimating must advance constructive suggestions to Accounting on how to improve them and work with Accounting to help obtain and install the improved rates. Senior management must give Estimating not only the right but also the *responsibility* to work jointly with Accounting.

Estimating's function is inextricably related to Accounting's function. This proposed joint responsibility for the costing and burden rates applies whether it is Estimating or Accounting that applies these rates. In either case, they result in the total estimated product cost, which is Estimating's responsibility. Estimators can't afford to apply blindly whatever rates they are given to use. They must understand and accept the logic with which the rates were developed and are to be applied. Experienced estimators, knowing as they do the product and processes, know the plant. They know how material, worker and/or machine hours, and overhead costs are collected and where and how they should be charged to individual work centers and eventually to individual products. This kind of knowledge of the operation, combined with Accounting's knowledge, should jointly produce better costing and burden rates than rates done by Accounting alone. The benefits apply to both functions:

The Controller will have the best product costs available from the talent on hand with which to value the inventories.

The Estimating function will have costing and burden rates which they understand, agree with, and have confidence in using.

The Estimating-Accounting Interface

Estimating depends on Accounting in, among others, two important areas: (1) Some or much of the costing information that Estimating uses comes from Accounting's records of past actual costs for similar work. It has been said that all estimates involve comparative evaluation of new requirements to analogous histories. (2) Because it is the department that collects actual costs, Accounting is the department that can supply the eventual actual costs for a job and thus provide the data for estimating follow-up.

Much has been said and written to the effect that in order to develop cost estimates in new products, Estimating may rely too much on past history of actual costs as represented by Accounting records. But in many situations, estimators have nothing else to use. In fact, for many estimators the problem is one of getting enough usable actual cost records. Often these records are not available or are not available in comparable usable form. Also, in many real-world estimating situations, estimators must work under tight time constraints. Marketing wants the estimate yesterday, or the prospective customer assures you that the competition already have their bids in and it needs yours *now*. There is not enough time for detailed research and study that estimators would like to do. They have to go with what they lay their hands on quickly, which often means using historical data. The use of such data is not wrong, but there is a threefold danger.

OVERRELIANCE ON HISTORY AS A PREDICTION OF FUTURE COSTS

Certainly, when we use such data we must be reasonably certain that they are usable in the new product or job. Using them blindly without analysis and thought is not adequate estimating. Also, it is natural when one has historical records to become overreliant on them. We must recognize when they are out of date and not fail to urge management to replace them with better data.

BLAND ACCEPTANCE OF PAST INEFFICIENCIES

Historical data reflect past actual efficiencies. Unfavorable variances from any standard that may have existed in the past are already built in and are included in past costs because they are actual costs. In using present cost standards we are applying past actual unfavorable or favorable variances to those standards, to develop the estimate. As a management function Estimating should indicate on the estimate any identifiable and continuing inefficiencies. Estimating may judge that these inefficiencies must be accepted on the new estimate, but Estimating should point out that these inefficiencies are included and not routinely build them in without comment.

AVOIDING THE NEEDED COST DETAIL

The historical cost data available may be in broad categories or groupings of costs. Under harsh time pressures it is only human for estimators to use the gross data they have and avoid the detailed job analysis that might be necessary for the new product or job. As a result, historical data may be used in cases where they really are not applicable.

Accepting the fact that Estimating must often use, in part at least, Accounting records as a data base from which to develop specific predictions, it becomes apparent that the company's cost records must meet certain minimum standards:

1. Stock-room issues of raw material and purchased components must be by type, cost, and job and accurately counted and reported.
2. Material use records must be by type, cost, and product or job and accurately reported.
3. Wherever possible, direct-labor reporting must be by work center, skill grade, product, and, if applicable, operation number. Such detail is not always feasible in practice, e.g., in multimachine operations or situations with no route sheets.
4. Production counts must be reasonably accurate. This is a real problem in some industrial situations. Strong management effort is needed to correct it when production counts have been allowed to be wrong and stay wrong.
5. Factory operating expenses must be reported under a well-defined, usable chart of accounts with charges to the specific work center whenever possible.
6. Factory overhead costs must be accumulated by, and charged to, the specific work centers or subcenters that incurred them. If they were incurred by a number of centers, they must be logically allocated to those centers individually.

These are the minimum requirements of cost accounting. In each real-life situation specific conditions often require other cost-reporting and collection action by Accounting to service their own and Estimating's needs. Additional requirements of cost accounting may be required because of the type of product and/or industry. For example:

In a highly engineered product, engineering costs should be kept by project and/or product line, depending upon the type of engineering.

Where distribution, sales, marketing, and advertising costs are high, such costs should be identifiable by product wherever practical.

These Accounting and shop-reporting requirements are all quite obvious and certainly easy enough to list, but they take a strong and determined management to obtain. In too many companies serious deficiencies in these areas are allowed to continue. Consider items 1 and 4 in the list above. Many companies have excessive discrepancies between book and actual inventory. Strong action is needed to close such gaps through more controlled and effective stock-room and shop-floor reporting. Many a Material Requirements Planning installation is in trouble today because the necessary discipline in material reporting in the stock rooms and on the shop floors has not been acknowledged and implemented. If shop counts or production counts are not as good as they ought to be, management has only itself to blame. Although in many shops, operators and supervisors are allowed to continue to report finished counts higher than the counts reported on previous operations, no one from Production Planning, Accounting, or any other management area raises questions and insists on corrective action. In contrast, in other companies such consistently poor shop reporting would never be tolerated.

These are all matters under management's jurisdiction. They can be controlled if the management is smart enough to recognize their importance and strong enough to see that they are corrected. Good reporting is achieved in too many companies to justify the belief that it cannot be accomplished.

From Estimating's standpoint the aim of this reporting and accounting procedure is to be able to develop and have available for use:

Good costing rates by worker hour or machine hour

Fixed or overhead costs that can be identified and charged to the product that incurred them if possible; otherwise allocated logically over the products to which they should be allocated

Usable data on past actual costs for potential use by Estimating in estimating the costs of similar future products

Follow-up data for estimating, by element of cost, to allow comparison with the costs originally estimated, making continued improvement in estimating possible

Problems in Overhead Allocation

It may be obvious that overhead or burden should be allocated properly to the different products made, but the facts of business life can frequently cause problems in such allocation. Let's look at a few examples.

• The general accounting format, the cost accounting structure as represented by the chart of accounts, and/or the cost-collecting procedures may not permit the accumulation of the cost data needed. For example, in the chart of accounts the cost account detail may be fine for General Accounting's use and profit and loss development but too gross and cumulated to provide the detail needed for good product costing. The data may fail to meet Estimating's and Production management's needs.

• Accounting and Estimating may not have the time to conduct analytical studies to determine the proper basis of allocation of fixed costs to individual product lines.

• Accounting and Estimating talent and motivation to make such an analysis may be lacking.

• Prejudice and emotion in senior management or Marketing and Sales may restrict the degree of realism attainable and applicable in the costing of the company's products. Odd as it may sound, I have seen managements actively resisting improved product costing because of the effect it would have on favorite products.

Like so many areas of management, overhead allocations and product costing as a whole are the art of the possible. No company will ever be completely satisfied with its product costing or its overhead allocations, but our goal in any company should be one of evolutionary improvement. That requires identification of our present weaknesses and errors and training or hiring the necessary talent. We must improve the procedures and the data developed as we can afford to. Certainly this year we want to be estimating better than we did 2 years ago.

Allocating Overhead Realistically

Good manufacturing cost estimates must be based on realistic costs. The key word is "realistic." We are concerned here with realism in overhead cost allocation. This requirement has two implications:

Burden costs arising from a given product must be assigned to that product and included in the costing rates used by Estimating for that product.

General burden, as well as specifically assignable burden, must be distributed and included in costing rates at forecasted volume to achieve the complete costing needed for pricing guidance.

A practicing estimator had an interesting reaction to the second point. "When you do it this way, if next year you have a lower forecasted sales volume than this year, your costs for next year will go up. And, if you have a higher forecasted sales volume, your costs will go down." The blunt answer to this is "Of course." If the company has lower forecasted sales and the forecast is realistic, it will have less income and operational activity over which to spread the fixed costs and the total cost per unit produced, including overhead, will be higher. Conversely, if a sales forecast higher than last year's actual sales is realized, there will be more operational activity and output over which to spread the fixed costs and in fact the total cost per unit will be lower. The observation, I think, indicates insufficient understanding of Accounting's problem and procedure. Accountants must have a volume base over which to spread the fixed costs or overhead. The only base they have is the best available sales forecast they can get their hands on. With that volume base they are able to develop with Estimating a total product cost, at least to the total manufacturing cost level. I say "to the total manufacturing cost level" because most companies charge selling and administrative costs monthly right into the profit and loss statement. The controller needs product total manufacturing cost to defend the asset of inventory, which is vital in producing a correct profit and loss statement and balance sheet. Again, the estimator has to understand accounting.

The proper allocation of fixed costs is a problem plaguing many managements and many Estimating departments in many companies. As we shall see in Chap. 5, in some companies Estimating does not even apply burden rates. In other companies Estimating blindly applies whatever burden rates Accounting provides. Many estimators believe that burden rates are none of their business, but I don't buy that. If Estimating has the responsibility for developing realistic predictions of a product's cost, and if overhead is a sizable percentage of that cost, Estimating must be involved in determining burden rates. As proposed earlier, the burden rates would be the *joint* output of Accounting and Estimating. I hold very strongly that Estimating must understand how Accounting has developed the burden rates it gives Estimating to use. Estimators must agree with the logic behind that development and with the application of that burden. If burden rates do not make sense to Estimating, Estimating should extend constructive suggestions on how the rates should be developed and what their content should be, by product and by cost center. Estimating should express its views not only to Accounting but also to Marketing and executive management if necessary.

The need for proper overhead allocations via good burden rates is so great that it deserves more attention than it seems to receive. More should be written on the subject. There should be seminars teaching it. The problem is that the techniques of developing proper fixed-cost allocations depend on the process, the product, and even the organization of the enterprise. What can be said on the subject? There are only a few basic, generally applicable fundamentals.

IDENTIFYING FIXED COSTS

We must first determine objectively which cost accounts are really fixed, at least for the present and forecasted activity ranges. This requires good cost analysis and identification (see Chapter 3).

ANALYZING FIXED-COST ACCOUNTS

Each of the fixed-cost accounts must be individually analyzed. Why is the cost incurred to begin with? Does it arise because of an outside influence like government regulations? Or because of a specific management function? Or because of the process itself? Next, possible effects or events under which that cost might rise or be controllable downward must be identified. All these details have to be considered and analyzed. Some fixed costs are justifiably assigned directly to a given product. More often a two-level allocation is required, in which certain fixed costs are first assigned to a stage of the process or a department in the organization and then allocated to products. In complex manufacturing situations more than two levels of allocation may be required before fixed costs are charged to the products themselves. It is usually in these multilevel fixed-cost-allocation situations that logic is weakest. For example, you have to point out that the present allocation obviously overcosts products A, B, and C and gives product D a partly free ride.

Unfortunately, at the present state of the art, when this analysis work is finished there still are huge amounts of fixed costs that you have no better way of applying to given products than on the basis of broad averages such as square feet, machine hours, or whatever. But at least, you won't be charging an uncrated item with crating costs, as the company did in Chapter 3.

APPLYING THE FACTS

Finally, the facts derived from the analysis must be applied: the proper fixed costs must be assigned to the right products. Surprising as it may seem, this is not always done. For example, better analysis of fixed costs can reveal the need for assigning more costs to given product lines. These higher cost assignments will have an obvious effect on the costs of those products and thus eventually their prices and marketing patterns. Marketing and Sales actively resists such cost reassignments, which therefore never get done. Executive management does not follow through, and poor product costing continues. These are situations in which management prejudice and emotion are allowed to prevail.

In my experience, it is in the area of realistic overhead allocations that estimating and cost accounting have the greatest need for good analytical thinking, sound logic, and a realistic approach. I always feel that one of the worst things that can happen to an estimator is to lay the estimate down before a senior executive, explaining how the estimate was developed and how the total costs were built up, and then have the executive prove that it is illogical to include certain costs assigned the product and included in the estimate. It is especially likely to happen in the overhead costs included in the estimate.

Results of Misallocated Overhead

Proper overhead allocation is important because of the great effect it has on the total product cost calculated and used. Important and long-lived management decisions are based on those costs. The most obvious wrong decisions that can result from misallocations of overhead would include the following:

1. Incorrect product costing and estimating
2. Incorrect pricing
3. Erroneous marketing policies and programs
4. Narrowed profit margins
5. Capital investment errors
6. Erroneous management decisions on product expansion and contraction
7. Wrong make-or-buy decisions

Accounting Requirements for Sound Estimating Follow-up

The observation that when the lessons of history are ignored we are condemned to repeat them applies to estimating. Since estimates are a prediction, it is essential to take the time and spend the money to check on the relationship between our prediction and the eventual actual cost. Only such facts will tell us where our estimates need improvement or where cost overruns occurred that may be corrected and avoided in the future.

Estimating invariably depends on Accounting for the actual cost data needed for estimating follow-up. In many companies today this follow-up job is poorly done or not done at all because Accounting lacks the staff and capacity to provide Estimating with the data in the necessary detail. But Accounting's capacity should be greatly improved with the advent of the computer. In fact, one hears of more and more cases in which Estimating still does not have the computer available but is getting new or better follow-up data because Accounting is now on the computer. Typical is a company making heavy equipment with $4 million a year in sales volume. The only computer on hand is a small one, newly installed, and its only output is for Accounting, using a purchased software package. Nevertheless, for the first time, Estimating will be receiving monthly actual material and labor costs versus estimated costs, by job and by major components or subassemblies of each job.

There are five requirements for adequate estimating follow-up.

• Cost Accounting records for direct or variable costs must be organized so that direct or variable costs are collected, if possible, by job or product. For example,

Material costs must be reported and collected by job or product.

Worker and/or machine hours must be collected by job or product.

Special supplies and tooling must be charged to the jobs or products that used them.

In practice, the usual situation is that the follow-up data collected and compared with estimates are these direct or variable costs, such as material, labor, expensive perishable tooling used only on a given job and charged against that job, special supplies, special operations done outside, etc. When the follow-up is on large, expensive equipment, the collection of actual costs for comparison with the estimate is by major component or subassembly of the final product. Such a breakdown allows better pinpointing in the follow-up analysis of actual versus estimated costs.

• Where it is possible and the money amounts are great enough, the cost structure for burden or overhead costs must be in enough burden-center detail to permit usably accurate cost collection against the specific job or product that incurred those costs.

• Actual costs must be collected, by job or product, and compared with the costs originally estimated in a programmed series of steps in specified time periods. For example, if the computer is available for use in this follow-up work, the original estimate data must be part of the input so that the output control or follow-up report can show the comparison between actual and estimated costs.

• Responsibility for the comparison of prediction with eventual fact must be assigned. Here again, we are in an area of joint effort by Estimating and Accounting. Certainly the controller's group wants to review the comparison to identify areas of cost overruns or Estimating undercosting. Similarly Estimating needs the comparison to take corrective action on future estimates.

• Most important, the findings of the comparison must be used. The lessons of history must be applied to correct estimating gaps or errors or to initiate the proper corrective action on excessive production or burden costs.

The procedures and details of estimating follow-up vary with each enterprise. Two examples will suffice.

• Where the product lines are fairly standard and quantities of production are high, a job-by-job comparison of estimated versus actual cost is impractical and unnecessary. In such cases a standard cost installation (the standards of which are used in Estimating) will show the amounts of volume variance, purchase-price variance, and controllable cost variances in material and labor costs. These variances are practical indicators of any gap between predicted and actual costs. However, when a new product line is initiated, a detailed comparison at the start between estimate and actual product cost is indicated. This may require temporary action for special cost collection and special detailed analysis, but the extra effort is good insurance and can detect errors in the initial estimate that need correcting.

• Where the manufacturing effort is on custom jobs, particularly jobs of high value, a detailed job-by-job comparison of estimate versus actual cost can be vital to survival.

The management art involved in such estimating follow-up is to determine the proper balance between the effort and cost of the follow-up and the results to be achieved from the knowledge gained and from the application of such knowledge. Our common experience in this matter should put us on guard:

Too many managements gamble on the overall and continuing accuracy of their company's estimating effort.

There is a reluctance to incur the cost of a good follow-up effort and a willingness to accept overall operating results as the measure of estimating accuracy.

Most companies could improve their profit performance by a more detailed analysis of estimated versus actual cost performance and by taking action on the findings of the analysis.

Relationship between Estimate and Price

For most commercial products, the chief determinant of price is the competitive market. For engineered products there tends to be a much closer causal relationship between estimated cost and price.

In every case, over the long run, the cost estimate must establish the *lower level* of the product price. Obviously, no enterprise can afford to ignore this lower level of price on too large a segment of its product line over too long a period of time. The result would be bankruptcy!

Unfortunately all too common is the situation where poor cost estimating and product costing result in an inaccurate cost, which in turn leads to an incorrect and often too low product price for certain products or product lines. These incorrectly costed and priced lines are supported by profitable lines, which may be just as inaccurately costed but whose market permits a higher price level. The result is the all too familiar one in which one or more products are carried by others. The result is a lower overall profit.

Good pricing and good marketing decisions depend on accurate cost estimates. This may appear self-evident, but it is a principle that often is imprecisely applied.

Again the importance of proper fixed cost or overhead allocations must be stressed. Such fixed costs are an important segment of the estimated product costs, in some industries being the major segment. In any case, estimators must help determine whether the allocations in use are reasonable. If not, estimators should take the lead in having them corrected.

Although executive management and Marketing are responsible for pricing, Estimating is responsible for giving them the facts. Executive management and Marketing may consider it necessary in the light of actual market conditions to ignore the cost facts provided by Estimating and follow a competitive price, but that does not absolve Estimating of its sometimes unpopular responsibility for collecting and presenting the facts.

Conventional Absorption Costing versus Direct Costing

Most companies today use *conventional absorption costing* (CAC) for their products. This method is decades old and has the great advantage of all those years of precedent and familiarity. It has a basic, readily understandable logic. It makes sense, particularly to the newcomer to business, to product costing, and to management. However, as we shall see, it also presents certain problems in management decision-making situations. *Direct costing* (DC), a different approach to product costing developed over the last 40 years, will probably be used more and more in the years ahead. Using DC has a great effect on how the estimate is structured. Estimators should be aware of DC, the logic behind the concept, its basic advantages and dangers, and how it affects the job of product-cost estimating.

CONVENTIONAL ABSORPTION COSTING

Since later we shall compare the two approaches to product costing, let us build a simple model of costing a product under CAC. Assume that the product costs $1 for material and $1 for direct labor. Also assume a 300 percent manufacturing overhead applied against labor. Finally, for simplicity, assume a 10 percent selling and administrative cost and a modest 10 percent markup for profit. Given these model parameters, CAC for the product would be as follows:

Material	$1.00
Labor	1.00
300% manufacturing overhead	3.00
Total manufacturing cost	$5.00
10% selling and administration	0.50
	$5.50
10% markup	0.55
Desired selling price	$6.05

This is the simplest model of CAC I can think of. It is a deliberate oversimplification since in real product costing more cost breakdowns would be used, the markups for material and labor would probably differ, and the costing would be more complex. In its essence, however, it is how you cost your products to arrive at calculated or desired selling prices if you are using CAC. Now let us cost this same product using the DC approach.

DIRECT COSTING

To do direct costing you must have analyzed and identified your costs (Chapter 3). That analytical work enables you to differentiate between (and thus segregate) your fixed and variable costs. On the above model, let us examine that $3.00 of manufacturing overhead cost. What does it consist of? It is actually an unidentified aggregate of fixed, variable, and mixed costs. Assume in this model that with good cost analysis you have determined that $0.75 of the $3.00 of manufacturing overhead is variable and the remaining $2.25 is fixed. Given the data now at hand, the direct costing for the model would be as follows:

Material	$1.00
Labor	1.00
Variable manufacturing overhead	0.75
Total variable costs	$2.75
Fixed manufacturing overhead	2.25
Total manufacturing cost	$5.00
10% selling and administration	0.50
	$5.50
10% markup	0.55
Desired selling price	$6.05

Let me hasten to add that this is a very simplified model of DC. For example, variable increments of cost would exist in real life in selling costs and in administrative costs; markups would probably be handled in a more sophisticated manner; there would be additional cost breakdowns. Also, with DC the development of the calculated or desired selling price would be approached in a different way, illustrated below. But again in essence, we have applied the DC approach to this product. Now we can contrast the two approaches.

	CAC	DC
Material	$1.00	$1.00
Labor	1.00	1.00
Variable manufacturing overhead		0.75
Total variable costs		$2.75
Fixed manufacturing overhead		2.25
300% manufacturing overhead	3.00	
Total manufacturing cost	$5.00	$5.00
10% selling and administration	0.50	0.50
	$5.50	$5.50
10% markup	0.55	0.55
Desired selling price	$6.05	$6.05

Apparently there are only minor differences between the two approaches. But consider. With DC we know a very useful and important fact we did not know with CAC. We now know that every time we sell one of these units we have

Sales volume	$6.05
Variable costs	−2.75
Contribution	$3.30

Every time we sell a unit at the desired selling price, out of the $6.05 of sales volume or income received, it will cost us $2.75 for direct or variable costs. We therefore have $3.30 left toward our fixed costs, and once past the break-even point this $3.30 will go completely into profit. (We do not make profit until we cover or pay for our fixed costs.) The *contribution margin* on this product is

$$\frac{\$3.30}{\$6.05} = 55\%$$

This is *not* the profit margin; it is the contribution margin that goes first toward covering the fixed costs and then, once past break-even, toward building up profit. Thus, the contribution margin is the percentage of what is left from sales income after the variable or direct costs are paid divided by the sales income. The important fact is that this contribution margin varies between products and models, between plants and divisions, and between customers, salespeople, and sales territories. When a management knows these contribution margins, in whatever breakdown is most useful in their given situation, it is able to make decisions that are much more informed and productive than ever possible with CAC. For this reason and for additional reasons (discussed under the advantages of DC) more companies will be using DC in the future. When a company goes to DC, it obviously affects how estimators structure the estimate. They must keep the variable costs segregated in order to predict not only the manufacturing and total cost but also the estimated contribution margin. All estimators I have met whose companies are on DC agree that its advent meant changes in their estimating approaches and structures.

The DC approach is new to many estimators. They may have heard it mentioned without any experience in its use. For this reason we give a brief review of other ramifications of DC.

• The contribution margin has also been referred to as the profit-volume (P/V) ratio. The two terms *contribution margin* and *P/V ratio* are synonymous. Actually, the "P/V" is a misnomner. It is not the profit-to-volume ratio or percentage. It is the percentage contribution to fixed costs and then, after break-even, to profit. The term *marginal contribution*, though longer, is more accurate than P/V ratio.

• In the simple model of DC above, the desired selling price was calculated in parallel to the CAC model, but in using DC data to calculate the desired selling price, the usual procedure followed is somewhat different. If the company knew the product's variable or direct cost per unit to be $2.75, and if their desired marginal contribution on that product was 55 percent (really 54.545 percent), the $2.75 total variable cost would be divided by the complement of 0.54545 to arrive at the $6.05 desired selling price. Thus

$$\frac{\$2.75}{1.00 - 0.54545} = \frac{\$2.75}{0.45455} = \$6.05$$

• As most readers know, in CAC in most companies a unit made and placed in inventory would be valued at its manufacturing cost of $5 only. As a result, $5 of cost and value would not show as a cost on the month's profit and loss statement but as a value in the asset of inventory on the balance sheet. Then when the unit was sold, it would be taken off that month's balance sheet and included as a cost on the same month's profit and loss statement. It is done this way because most companies charge all selling and administrative costs, as costs, on the monthly profit and loss statement and capitalize costs in inventory only at the manufacturing-cost value. In contrast, under DC, the company would place that item in inventory at only its direct or variable-cost value of $2.75 and take it out of inventory at that same value.

• Thoughtful readers pondering the implications of the previous paragraph will realize that this difference between the two approaches will have a great effect on the profit and loss statement whenever inventory levels at the end of the month are different from those at the start. When inventory increases, the CAC profit and loss statement will declare a higher profit than the DC profit and loss statement because costs of a greater amount ($5.00 instead of $2.75 per unit) are treated not as costs but as assets and because stating lower costs results in a higher stated profit. As a result, under DC, profit is tied directly to sales income. As sales increase, profits increase; as sales decrease, profits decrease, all other things remaining stable. For that reason, DC profit and loss statements are often referred to as marketing-oriented, whereas CAC profit and loss statements might be considered production-oriented. Conversely, of course, when inventory decreases, the profit and loss statement under DC will show a higher profit than the CAC profit and loss statement.

This means that DC is generally not popular with the Internal Revenue Service because lower declared taxes mean lower tax liability and payments. Thanks to sufficient years of precedent and/or local rulings, however, some companies are completely on DC. In most companies using DC, internal reporting is done on a DC basis. Then four times a year, for tax-reporting and stockholder-reporting purposes, the profit and loss, inventory value, and balance sheet are calculated on the CAC basis. Accounting makes the necessary conversions quarterly, and management has DC data available every month for internal use.

Problems with Conventional Absorption Costing

CAC is a well-established, accepted approach offering the great advantages of readily understandable logic and years of familiarity. But as management becomes more knowledgeable and more anxious to obtain data upon which to base better decisions, two basic problems with this concept become important: (1) The product-cost data obtainable with CAC are not good enough or detailed enough for some of the hard decisions that have to be made. One aspect of this problem is discussed below in the area of marginal pricing. (2) With only CAC data to use, managements end up making some illogical, self-destructive decisions. A single-minded devotion to CAC often results in misapplication of product costs and the consequent wrong management action.

Marginal pricing will be discussed in the next section. For the moment, let us consider the second point. The company or division has a product with a variable manufacturing cost of $5.00 each. Fixed costs assigned to this product are $25,000 a year. For model simplicity, assume a 10 percent selling and administration burden and a 10 percent markup for profit. The sales forecast is that we shall sell 10,000 units. Given these data, we can calculate a selling price for this product:

Variable manufacturing cost	$5.00
Fixed manufacturing cost ($25,000/10,000 units)	2.50
Total manufacturing cost	$7.50
10% selling and administration	0.75
Total cost	$8.25
10% markup	0.83
Selling price	$9.08

Assume that the product is attractive, the price is right, and we end up the year selling not 10,000 units as originally forecast but 12,000 units. Under CAC the logic in pricing for the ensuing year can go like this:

Variable manufacturing cost	$5.00
Fixed manufacturing cost ($25,000/12,000 units)	2.08
Total manufacturing cost	$7.08
10% selling and administration	0.71
Total cost	$7.79
10% markup	0.78
Selling price	$8.57

In other words, the product is a winner and the price is right. Cut the price? That might not be bad pricing strategy if the lower price will sell 16,000 units next year. But the more curious logic is seen in the other direction. Suppose the product is not quite as attractive as we hoped, and/or the $9.08 price was not as attractive, and we end up the year having sold only 8000 units. Now under CAC the pricing logic for the coming year can go like this:

Variable manufacturing cost	$5.00
Fixed manufacturing cost ($25,000/8000 units)	3.13
Total manufacturing cost	$8.13
10% selling and administration	0.81
Total cost	$8.94
10% markup	0.89
Selling price	$9.83

Sales are down. Raise the price? A curious logic. It might be called self-destructive. You might not think this goes on, but it does. I have often witnessed it when a company has a broad product line and segments of the fixed costs have been assigned to each product. When sales are down one for or more of the products, the reaction is: We have to get more money for those products. Studies by independent institutes and industrial boards have found that much pricing practice follows this pattern. Many managers assure me that they have observed it at work in their own companies.

Consider a second example of the kind of curious and unfortunate logic a management can be led to follow by single-minded devotion to CAC. This example of make-or-buy decision making is an actual one which readers are asked to keep in mind for discussion later in the book.

I was being given a tour of a large electronics plant in the Midwest. The company makes sophisticated electronic black boxes, many for defense orders, and was interested in a consulting proposal. The executive taking me around was in the higher levels of middle management and had just returned from abroad, where he had helped negotiate a large defense contract with a foreign government. He impressed me as very knowing and down-to-earth. In the course of our tour, we came to the Transformer department, a large production area, full of equipment, with only two or three people at work. I asked my guide, "You must be using a lot of different transformers in the products you're making. Where is everybody? It's not coffee break or lunchtime." His rueful reply was, "It's a sad story. Some years ago, our controller's group made a make-or-buy study on one of our largest-volume transformer items. At our existing hourly labor cost including manufacturing overhead it was obvious that this transformer would be cheaper to buy than to continue making. So, we bought it. As a result, we had less

labor in the department and our cost per labor hour, including manufacturing overhead, went up. This turned other present "makes" into
obvious "buys" until we gradually worked our way into what you see
here today." What incredible logic!

Marginal Pricing with Direct Costing

Many Marketing or Sales managers till hard soil. They are under
constant competitive pressures to reduce prices. With DC and its
marginal-contribution facts, the data are available to indicate how far
managers can go in accepting price reductions. If a product has a $50
selling price and its direct or variable costs are $30, there is a $20 range
within which prices can be adjusted downward before contribution
reaches zero. But good product estimating is essential for its use.

In real-life marketing there are many instances where marginal pricing can or must be considered:

The profitable use of open capacity

Retaining existing business in the face of strong price competition

Breaking into new markets or finding new customers for existing
products

Consider one happy example of marginal pricing. Your company is a
famous silverware manufacturer making one of the nation's finest lines
of sterling and plated flat- and hollow ware. Your sales outlets are the
best department stores and jewelry stores across the country, and you
will do nothing to harm your marketing pattern. You make a nice profit
but have unused capacity. A major breakfast cereal company approaches you with a proposal. They have a great idea for a $3 premium
item for the back of their giant cornflake box. They want to offer a
silver-plated miniature Revere bowl vase. They are old hands at such
premiums; their plan is to keep it on the box for 6 months, and the
forecast is to sell 1 million units. You will ship to them in bulk; they
will need $0.75 a unit for packing and mailing while you would receive
$2.25 a unit. Your estimators go to work and project that for the copper
metal, spinning the vase, stamping the base, soldering the two together,
cleanup and light silver plating the total direct or variable costs would
be $2.00 per unit. The spread between $2.00 and $2.25 is $0.25; multiplied by 1 million units that is $250,000. You are already covering your
total fixed cost and making a profit. This job will not disturb your
present marketing patterns or outlets. Therefore that $250,000 will be
an addition to profit. But the estimated $2 variable cost per unit had
better be right.

Danger of Direct Costing

Obviously marginal pricing must be used with great care and discretion. It can lead to an excess of low-margin and even underpriced sales. It can upset existing marketing relationships with customers if used unwisely. Like any technique, it has dangers and creates problems if misused. For example, consider the first DC model given above. In some companies Marketing or Sales, knowing the $6.05 desired selling price and the $2.75 total variable costs, have thought they had $3.30 to play with. When they reduce the selling price below $5.50 or even $5.00, the company is in trouble. You and I can only be aghast at the lack of pricing discipline that must exist in such companies. But they have even abandoned DC, which is like throwing out your table saw because you cut yourself. It is not the saw's fault.

I had one controller tell me, "We have direct costing, but we don't tell Sales."

DC is a sharp and potent tool that allows calculated marginal pricing where competitive conditions demand such pricing, but it must be used under disciplined senior management guidelines carefully followed by Marketing or Sales.

The Tolling Principle

An allied matter in pricing with DC information is the *tolling principle,* or sharing with the customer the additional profits arising from overabsorbed burden because of additional business from that customer. It is a tool of negotiation that requires good product estimating and costing on the part of the vendor.

For example, a substantial part of your total sales volume is with a given customer. You make a fair profit on its work but have the capacity to take on more work, and the customer knows all this. The customer's approach is that if it gives you additional orders, you will be overabsorbing your fixed costs. It asks how much you will toll back in the form of lower prices or a year-end payment which will have the same effect as lowered prices.

The tolling principle usually is used in large-dollar-volume purchases and in situations where a single customer buys a large percentage of total output. Naturally, it is usually the customer who wishes to apply the principle. When the demand comes from an important customer who offers a larger percentage of its business, it can become a meaningful basis for price negotiation. It is tough vendor negotiating.

In use, the principle demands good cost estimating and good burden costing and allocation to the product or products in question. Otherwise the application of the principle could obviously result in underpricing. Applying the principle may also entail divulging cost information normally not provided a customer. An alternative, avoiding such disclosure, is to have an independent outsider determine how much is due the customer under the tolling principle.

Advantages of Direct Costing

In a book on product-cost estimating the marginal-pricing application of DC and its demands on Estimating are the matters of prime interest, but that is only one use of DC. Here are a few other important uses and advantages of DC.

• You know the differing marginal contributions by product, by model, by plant, by division, by market area, by customer, by salesman, or by whatever breakdown management finds most useful. This is, in my experience, the greatest and most productive use of DC. With DC facts management can make more informed decisions on such matters as the following:

1. With given production capacity limits, what product items should business be concentrated on to achieve the maximum profits and return on investment?
2. With bottleneck equipment of limited capacity, which products and/or orders should be made on that equipment to yield the highest profits?
3. With excess capacity, what business should be sought and at what price can it be accepted?
4. What effect will a price and volume change have on profits, the break-even point, and margins of safety?
5. Where should the company's limited sales capacity be concentrated for the highest profit? Which products should be pushed?
6. What additional sales are needed to pay for added promotion?

• The DC approach emphasizes fixed costs and their effects. Its profit and loss statement is a harsh one because any month in which sales income does not cover the total fixed costs in manufacturing, selling, and administration will be specified as an unprofitable month.

• It presents an easy and clear way to develop break-even points and margins of safety.

• It fits management thinking more closely, particularly the Engineering, Marketing, and Production management mind, because the DC profit and loss statement is undistorted by under- and overabsorbed fixed costs.

Other more subtle advantages of DC need not be dealt with in this book. The greatest testimonial to this product-costing approach is one I have received at the management seminars. I cover the subject in the estimating, the marketing, and the budgeting seminars. I ask, "Whose company is using direct costs?" Of those responding affirmatively I ask, "Have you ever worked for companies using conventional absorption costing?" Of those who have had hands-on experience with both approaches I ask, "If you owned the company, which would you use?" To date without exception they answer, "Direct costing."

The odds are good that in the years ahead more and more companies will be taking the DC approach. When they do, more and more estimators will be structuring their estimates for DC and structuring them differently from the way they developed cost predictions under CAC.

Chapter 5

The Steps and Areas of Product-Cost Estimating

Whether the company is making biscuits, fountain pens, pumps, or airframes, nine basic steps are involved in developing the product-cost estimate. Beyond these basic steps, many estimators must consider such additional cost factors as engineering, tooling, extra physical facilities, etc. The basic estimating steps and additional cost areas are the subject of this chapter.

Basic Estimating Steps

No matter what the product, any cost estimate should involve the nine basic steps listed below. Not every estimator carries out all these steps. In some companies Estimating completes only the first five steps. In other companies, Estimating carries on through step 7. In a few companies Estimating completes all nine steps. The ninth step, specifically, can be a controversial suggestion to some managements, but it is included for a very good reason, discussed later in the chapter. The basic steps of estimating are:

1. Determine what is being estimated
2. Break down into parts list
3. Determine material costs

4. Route individual parts

5. Estimate operation and setup times

6. Apply labor and manufacturing-overhead rates

7. Calculate total manufacturing cost

8. Apply selling and general administrative burdens

9. Apply markups and develop standard selling price

Fundamental Responsibility of the Estimating Department

The content and responsibility of estimating was discussed in Chapters 1 and 2, but the matter is so important and so frequently misunderstood and not clearly assigned that it warrants further consideration before discussing the nine steps in detail.

It is not at all uncommon to see a company using a *conference method* of estimating, under which Estimating collects:

Engineering time estimates from Engineering

Material costs from Purchasing

Tooling costs from Tool Engineering or Manufacturing Engineering

Operational labor times from Industrial Engineering

Estimating then adds these various cost factors together to develop a predicted product cost. Thus each of the individual departments is responsible for its particular cost segment of the estimate. Estimating merely collects these data and adds them all up. In such a situation Estimating is a clerical function, *not* responsible to senior management for the development of an *independent, realistic prediction* of what it will cost to make the product. In turn, when senior management, or Marketing management wants to follow up to see where an estimate may have gone wrong by comparing it with eventual actual costs, they have to look at each cost segment and each department that originated that portion of the estimate to determine what specific estimated costs went wrong. This is laborious. Where estimating follow-up data are poor, late, or nonexistent, the chore is so difficult that frequently in real life it just doesn't get done. As a result, the company keeps repeating its estimating errors.

I am not suggesting that Estimating should not obtain estimates of specific costs from other functions, but I firmly believe that Estimating must be responsible to management for the quality, i.e., independence and realism, of the final estimate. They can obtain engineering times from Engineering, material costs from Purchasing, etc., but Estimating should have not only the right but also the obligation to question, if they believe it necessary, the projected costs supplied by other management areas. Here are three examples.

• If Estimating evaluates projected engineering hours as being too high or too unrealistically low in the face of actual engineering costs on past jobs, Estimating must raise the questions not only with Engineering but, if they do not listen, with senior management as well.

• If Estimating considers, say, the material costs supplied by Purchasing to be too high and likely to be reducible by further shopping on Purchasing's part, Estimating should ask for such action. Unfortunately, enough time may not be available. Then what is learned on this job should be applied to future jobs.

• If on the basis of past experience Estimating considers Industrial Engineering's estimated labor times to be unrealistically optimistic and/or incomplete, their responsibility is to raise the point emphatically.

In sum, it is all a matter of what we consider the responsibility of Estimating to be. What I am proposing is standard operating procedure in some companies. In many others, it is not. It demands a lot of Estimating. In some companies, for example, it would be politically hazardous for Estimating to question the engineering costs advanced by Engineering management or to ask Purchasing to be more aggressive in their shopping. But it is all the hard chore of the Estimating function *if* top management is to be able to look to Estimating for an independent, realistic prediction of what it will cost to make the product.

In some highly complex product-cost estimating situations, the estimates are for custom jobs, for one one-offs, or for relatively limited quantities of a high-value product. In such cases, after the estimator has estimated material and labor costs, Engineering must estimate the engineering costs involved because Estimating lacks the technical expertise to evaluate this important cost. Thus the final estimating responsibility has to be divided. This situation is too common to be ignored, but even here Estimating should do the follow-up reporting on the whole estimate and report on all unfavorable variances, including Engineering's, between estimated and actual costs.

The estimating steps discussed below are well known to any experienced estimator, but the discussion also deals with the extent or total content of the estimating job and with how, in many estimating situations, the estimating responsibility stops short of the total job. Some of the basic points that follow are often overlooked in many companies.

Step 1: Determine What Is Being Estimated

To develop an accurate estimate, estimators must know the specifics of the product for which the estimate is needed. If they do not have these specifics, assumptions about the product content have to be made. As explained in Chapter 1, in that unhappy event estimators should list those assumptions on the estimate sheet(s) together with a statement of the margin of error should those assumptions prove wrong when the product is made.

When you consider how many estimating situations you encounter in real life, you can understand why the spectrum of estimating conditions is so broad and varies so widely between different industries and products. At the easier end of the spectrum are estimating situations where the product is relatively simple, with relatively few parts. Typical firms include printing houses, box shops, most job machine shops producing parts to customers' prints, and simpler consumer items, such as toiletries and some appliances. Often estimators have available for use a specific bill of material or parts list, developed by Engineering, and detail drawings of the parts needed. Then an estimator can get right into a detailed estimate.

At the other, more difficult end of the spectrum is the situation where estimators have only a minimum of product specifics available. They are provided only with certain broad product characteristics or desired performances such as capacity, hourly rates of output, limits of tolerances, or whatever. Such situations are frequently and typically found in industries and companies producing heavy custom equipment, advanced machine tools, state-of-the-art electronics, etc. These situations, in which estimators must do conceptual (or parametric, or budgetary) estimating, arise in circumstances like the following.

• Marketing sees an opportunity in the marketplace. If the capacity or performance of the existing product line can be extended or the tolerance performance tightened, there is a possible market. They have a range, often narrow, of potential sales prices, for which such an improved product could be sold. The questions that need answering are: How much would it cost to produce such a higher-performance model of the present product line? What additional investment cost, if any, will be needed to produce it? Then, Marketing and top management can decide whether the market is worth entering or trying to develop.

• A good or potentially good customer needs a new piece of equipment or a new component for its final product. At this point the customer cannot specify anything but the operating or performance characteristics. No detailed bills of material—much less parts prints—have been developed and if your company would normally do this work, the customer will not allow enough time for all the estimating work involved. It wants a bid now, and your company wants to provide it with an estimated cost of the item in question.

• Engineering has an idea which advances the state of the art in the industry and which would require a product more advanced and/or of a different type from any the company presently produces. To determine whether such a product represents a viable idea, the cost to produce such a product must be estimated.

This last situation is, of course, not at all uncommon. It is how companies stay competitive and grow. Frequently in such situations, you will find Research and Development or Design Engineering doing the product-cost estimate. You will also frequently find those cost estimates turning out to be optimistic and unrealistic. Product-cost estimating has a place here. If the company has a good Estimating department, responsible for product-cost estimating, that department should work with Engineering to develop the conceptual cost estimate. Such a joint effort should increase the chances of a realistic estimate's being made.

Between these two estimating extremes you find an infinite variety of conditions. Even in a given industry practice varies widely in how much product detail estimators will have to work with. It all depends on the given management's philosophy and the subsequent practice. In some companies, senior management requires of Engineering only the broadest specifications and inputs; and as a result, the estimate, whether done by Engineering or Estimating or jointly, must represent very rough approximations and the potential error, either too high or too low, is great. It is the old "ice-water-in-the-veins" situation. In other companies, senior management requires much more specific and detailed product specifications and definition before using the product-cost estimate for decision making. If the customer is providing these specifications and will not or cannot provide them in sufficient detail, management will refuse to submit a bid or will amply protect the company by a detailed set of exceptions in the bid submitted.

Step 2: Break Down into Parts List

For an accurate estimate you need a complete bill of material, i.e., parts listing and subassembly sequencing for the product to be estimated. Again the range of actual estimating situations and practice is tremendously broad.

In a product consisting of, say, five parts, the bill of material is simple, and the possibilities of omissions or errors relatively few. In contrast, for such complex products as airframes, avionics, or heavy custom equipment the bills of material are huge. In such situations, an early step in these companies has been to put the bills of materials onto the computer to make them easy to retrieve. It is difficult for younger estimators in such companies and product lines to conceive how such enormous bills of material were handled before the advent of the computer.

In situations demanding conceptualized estimating, no complete bills of material are available for use by estimators, who instead must use whatever broad parameters of product characteristics they can. Weight, production rates, numbers of components, size, and tolerances are among the variables affecting product cost on which estimators may have available data. They may also have developed their own predictors of cost. Fortunately, in many such cases, the situation is not this bleak. The advanced model or the large complex custom item on the request for quotation will have some or many subassemblies that have been used in the past on other models or other custom-built jobs. In that event at least some of the bill of material will be available to estimators.

Step 3: Determine Material Costs

A complete estimate must include the cost of the three types of material used in the finished product:

1. Purchased components, if any, used in the final product
2. The direct material, such as bar stock, sheet, plate, chemicals or whatever, to which value will be added to make the parts used in the final product
3. The direct packaging material (corrugated box or chipboard box, crate, pallet, etc.) used to ship the final product

In estimating situations involving relatively simple products most estimators usually do such material costing and will therefore have past material costs to use. Or if new components or materials are required for the new job or product being estimated, suppliers' price lists will be consulted. A variant of this procedure can be found in situations involving relatively simple product but very high volumes. In such instances, where a difference of 0.5 or even 0.1 cent can be very important, estimators may well have Purchasing seek two or three bids or have Purchasing send requests for quotations and then negotiate the responses from a number of suppliers. Basically, in such cases, Purchasing determines material costs.

On products of great complexity and many parts, estimating situations and practice vary widely and depend on such factors as the time available to make the estimate, the nature of the product, the percentage that material represents of the total cost of sale, and the degree of detail available about what is being estimated. Here are some typical situations and practice.

• The product being estimated is complex, and each job is custom-designed to some degree, varying with the job. Fortunately in such cases, a great deal of the material and many of the components are the same as, or very similar to, those used on past jobs. Estimators can therefore refer to past actual material costs on similar past jobs. For many of the major purchased components, estimators can estimate reasonably well on the basis of the past costs of such specific components. In such situations estimators usually do all the material-cost estimating, particularly where time to get the estimate out is limited.

• On highly complex defense products, the rules of the game frequently require three bids for major purchased components or subassemblies going into the prime contractor's final product. There are many requests for quotations and negotiating sessions on the responses. Purchasing and Estimating work hand-in-hand to develop projected material costs. The procedure is lengthy. In highly complex, nondefense work estimators frequently do not have such luxuries of time and must develop their own material-cost projections.

• The most difficult and most unfortunate situation in material-cost estimating is the conceptual estimating where no (or only a few) bills of material and component specification data are available. In such situations the poor estimators are invariably on their own because they have too few data to give Purchasing to enlist their help. The only recourse is to use actual cost history on similar work, if any such data or jobs are available, and/or broad cost parameters such as weight, projected performance characteristics, estimated number of components, or whatever, of the new product being estimated.

We can list six chief dangers in estimating material costs:

1. Inadequate product specifications may cause the use of prices for parts or material that are below or above the correct quality level.
2. Incorrect or changing delivery requirements or vendor delivery failures may cause the need for more expensive substitutes or production changes that require additional setups or production delays.
3. Incomplete product specifications may result in material estimates

that do not cover the material costs involved in the final engineering design or product quality standards.

4. Material price levels may change upward and exceed estimated material costs, particularly in estimating material costs for products with long manufacturing lead times. In such cases, some companies and estimators tie their quote to specific price indices (see the section Projecting Inflation's Effects, below).

5. Material price breaks anticipated and planned on may not be realized because of such factors as delivery schedule changes, revisions to inventory policy, etc.

6. Estimates of material costs do not include the percentage additive necessary for special handling (say, air freight), extra inspection, etc., that may eventually be needed but not anticipated by the estimator.

If your company has had one or more of these problems in the past and continues to have them, you as estimators must use this past experience in developing new estimates. For realistic estimates these extra costs have to be built into new estimates; even better, Estimating should incite management to take corrective action.

From the estimators' standpoint, one further point about material costs should be mentioned. Estimators should not be content merely to reflect in their estimates projected material costs based on data provided by Purchasing and/or Accounting and their own analysis and judgment. They should also consider as part of their responsibility detecting areas of material-cost improvement and developing constructive suggestions to improve material costs.

These opportunities can take a myriad of forms, depending on the circumstances. For example, estimators may see:

Overspecifications of material by Engineering

The possibility of obtaining lower-cost substitutions from Engineering or finding less expensive sources by Purchasing

The potential for scrap reduction by Production

A quality estimating job is not done by simply reflecting past experience. Good estimators see their job as much more and should have the position and status to feel free to make constructive, practical suggestions. In fact, management should require them.

Projecting Inflation's Effects

We digress for a moment from the sequence of estimating steps to discuss a problem prevalent today and almost certainly in the future, that of estimating the cost effects of inflation. This problem is vital for products with long manufacturing lead time and/or long delivery times since the customer is ordering way ahead and taking delivery far in the future.

The question was well posed by the practicing estimator who told me that he was estimating costs on heavy power generating equipment with deliveries 5 years hence. An important material-cost item was heavy steel forgings that would not be bought and machined until 4 years hence. As he put it, "How do I know what steel forgings will cost 4 years from now?" His solution was to build into his costs and bid the current cost of the steel forgings needed on the job and also build into the contract an escalation factor for that cost based on the government price index for steel forgings 4 years hence or 1 year before delivery. I can think of no better approach in that situation.

For such long lead-time products, producers must protect their profit margins by building in some measure of inflation's effects. The key word is "measure." The cost-increase factor used must be quantifiably measurable and realistic to be competitive and fair to the customer. In many instances the *Producer's Index* (formerly called the Bureau of Labor Statistics Indices) is an adequate measure. In other cases it is not specific enough. In such instances you should develop and maintain your own cost records and develop the specific indexes needed. This way is less popular with the usual customer, who naturally will consider the government indexes more dispassionate and independent. Wherever realistic and usable government indexes are applicable, they are preferable to private ones.

In some industries with harsh competitive conditions, a firm bid is demanded if you are to get the order and the inflationary risks must be borne by the vendor. The happier extreme is where the vendor can quote "price at the time of delivery." In many cases the practice lies somewhere between these two extremes. I also detect a great difference in practice in this matter within a given industry, depending on the approach of the customer's purchasing managers. Some demand and get firm bids while others will accept price at the time of delivery.

Estimating Tooling Costs

Tooling in one form or another is usually required in the manufacture of a product. Metalworking requires cutting tools, fixtures, dies, and jigs. Foundries need patterns and core boxes. Die casting needs molds, as does plastics molding. Airframes need assembly fixtures. For good estimating, the costs of such tooling obviously must be included in the product cost estimate. Such tooling is of two types.

PERISHABLE TOOLING

This tooling is usually of lower unit cost, though in the aggregate it may amount to a very large cost item. Since it is tooling usable on a wide variety of jobs and operations, it is not chargeable to specific parts and thus is omitted from the cost estimate for those parts. Instead the cost of perishable tooling (drills, taps, dies, carbide or high-speed steel-turning tools, common-sized milling cutters, etc.) is included in the manufacturing overhead rates and thus in the product cost estimates. Two cautions:

• Where a given job or product requires exceptional perishable-tool consumption or expensive special perishable tooling (e.g., in metalworking, special form tools or broaching tools), such perishable-tooling cost is best charged to that job or product instead of being included in an overall burden rate. To do otherwise will undercost the job and have other jobs overcharged to carry it.

• Perishable-tooling costs that are appreciably higher for a given machine center or centers should be included in that center's manufacturing overhead or burden rate. This requires that perishable tooling be requisitioned and the cost of the requisitioned tools charged to the department or work center originating the requisition. Then the controller must use these collected actual-tool-cost data to develop cost-center burden rates. This is just another example of why good cost-center burden rates, not overall plant burden rates, are essential to good product estimating and costing.

I usually find costs of perishable tooling, particularly special-purpose perishable tooling, being estimated by Estimating using tool vendors' price lists. Much less commonly are Purchasing's inputs sought by Estimating.

DURABLE TOOLING

This type of tooling is used for a specific part or subassembly on repeated production runs (e.g., assembly fixtures, punch-press dies, plastics molding and die-casting molds, drilling jigs and fixtures, and foundry molds). In almost every company such expensive items are capitalized and their acquisition costs depreciated over the number of units expected to be produced from them. In some enterprises, some of the more expensive specialized tooling (like forming tools, broaches, or hobbing cutters) that other companies would treat as perishable tooling will be handled as durable tooling and capitalized. In some industries the reverse is true, trade custom being such that expensive mold

equipment that would be capitalized in other industries has historically been expensed and the precedent applies throughout the industry. Obviously the differentiation between durable and perishable tooling is not clear-cut and varies between industries and even between
companies within a given industry.

Durable-tool maintenance costs, such as die or mold repair or tool
sharpening, is almost never capitalized. In 35 years of consulting practice I have never heard of an instance where such costs were not
expensed and thus picked up by the burden factors applied in the
estimate.

Good product-cost estimating requires that expensive durable tools
required to make the product be charged to that product and their costs
not spread over the cost of other products. To do otherwise is to undercost that product and overcost the others that do not require that tooling. It is all so obvious to write and say, but frequently in real life the
principle is ignored.

In some estimating situations the cost of such durable tooling must
be estimated in product-cost estimating. In other companies, Tool Engineering or Production Engineering does this for Estimating. I have
always believed that where durable costs are an important part of manufacturing costs, Estimating itself should develop the talent and ability
to establish reasonable and practical tool-design time and building-
time costs and thus develop the durable-tool cost estimate. Then
whether the tool is built in the company's own toolroom or bought from
an outside tool and die house, the company has an independently established cost standard against which to compare actual durable-tool cost.
In any case, no matter who develops the durable-tool-cost estimate, it is
a wise management that keeps a continuing follow-up check on actual
durable-tool cost versus estimated cost. My experience has been that
only a few companies do this. Frequently actual costs exceed estimates,
and as a result the product is undercosted. This is an area difficult to
control. For example, very few companies have toolroom standards.
Much remains to be done in this area.

Next, in developing a product unit cost estimators have to spread the
estimated durable-tool cost over a given production quantity. Estimating practice on this matter varies widely between industries and between companies within a given industry. In some situations, estimators are given firm guidelines established by the controller, based
in some cases on marketing forecasts of the number of units that will be
sold. In other situations estimators can be said to have *carte blanche* to
decide on their own how many units the durable-tool acquisition cost is
to be spread over. The whole matter is a "between-a-rock-and-a-
whirlpool" situation. If the quantity of units selected over which to

spread the tool costs is very large, the estimated costs will be lower but the company may not sell that many and then the costs are underestimated. If a more conservative approach is followed and a low number of units is chosen, the estimated costs per unit are necessarily higher and thus quite possibly less competitive. I have always judged it best, if error is made, to err on the conservative side and try wherever possible to spread durable-tooling costs over a conservative number of units to be made. While this results in a higher estimated product cost, it furnishes attractive insurance against earlier than anticipated product obsolescence, possible future product-engineering changes, and unexpected shorter durable-tool life or accident to the tool.

Many product-cost estimators who must also estimate durable-tooling costs find themselves quite alone. Very little is published on the subject. They have no durable-tool cost standards and usually have to resort to actual durable-tool costs on tools that may not be close enough in nature to the durable tools they are now estimating. They can use some help and are referred to the Bibliography. The Leonard Nelson article in particular has proved helpful to a number of practicing estimators I have met. There is much more to be learned by research and visits within your own industry. It is entirely possible that other estimators in other companies in your industry or in allied industries have developed approaches and even data that might be made available to help you. You never know if you don't ask.

Estimating Facility Costs

In some situations, the product or job that is being estimated will require additional or modified production and/or plant facilities, e.g.,

Additional power, air, or storage facilities may be required if the job or product is to be built.

The job may require additional floor space to be cleared for its erection.

Added maintenance facilities may be needed.

If additional production and/or plant facilities are needed for a specific job or product, their costs should be charged to the product or job and thus must be included in the product-cost estimate.

In some instances estimating the cost of new facilities is so specialized that Plant Engineering talents and inputs must be called upon by Estimating. However, it is frequently advisable to have experienced product-cost Estimating personnel involved in the development of the final cost estimates of the additional facilities, if only on a review basis.

Step 4: Route Individual Parts

Routing is the determination and listing of the sequential operations required to make the given part or assembly. Normally, it includes not only production steps but inspection points as well. The operations are listed on variously named route, process, or operation sheets.

In the usual product-cost estimating situation, the product-cost estimator must do the initial routing. Many estimators on custom work in metalworking have to work from parts sketches or prints, with only the notation to "machine complete," or "machine as per blueprint." The final detail routing by Manufacturing or by Industrial Engineering will be done only if the job is actually sold. In estimating new proprietary products expected to become part of the company's continuing product line, another department such as Manufacturing Engineering may provide the estimator with the parts routing. For complex assembly work, the bill of material made up by Design Engineering should indicate the subassembly sequence. Frequently in these situations, however, Estimating must also check with the assembly departments or erection shops to see whether the actual subassembly sequence and parts contents of given subassemblies deviate from the way Engineering laid them out. The practice varies between different companies and industries and often depends on in-company precedent. In any case, Estimating bears the responsibility for the realism and accuracy of the routing used for parts and assemblies, because Estimating is responsible for the final product-cost Estimate.

In some industries and products operation routing is not needed at all. The material can only go to one next operation after the previous operation is completed. Most chemical production is an example of this situation. In other industries the routing is simple enough to be indicated on the shop order, and no separate route sheet is needed, as in corrugated-box manufacture. In contrast, in producing complex metal parts or in avionics assembly extensive and detailed route sheets or operation sheets, listing 20 to 40 or more operations, are needed.

In conceptual estimating, usually no parts sketches, much less parts blueprints, are available and routing is not possible. If parts and/or subassemblies from previous products are to be used in the new product or job being estimated, however, routings should be available for them and the estimator should find and use them.

Step 5: Estimate Operation and Setup Times

After the operational sequence has been established, the estimator must predict the time needed to perform the individual operations. If setup work is needed on certain operations, setup times must also be estimated.

As already noted, the prime reason to have and use engineered labor standards is for labor-cost control. Such labor standards are then available for use in product-cost estimating and product costing, make-or-buy calculations, capital investment decision making, and all the other management matters that require labor time standards.

Estimators who do not have engineered labor standards to use must use data from Accounting. If they do not even have decent actual data, they must use their last resource—their own or production supervision's best guess. They have nothing else to use. In that position, however, estimators should urge management to spend the money to develop usable labor standards or at least restructure their accounting procedures so that decent historical records are kept.

Estimators without either engineered labor standards or adequate historical records should research what is available from outside their company in the industry. For example, in metalworking, comprehensive machining standards can be purchased (see the Bibliography). In other instances, usable standards can be obtained from the machinery manufacturers and/or the supply manufacturers, e.g., carbide-tool manufacturers and welding-rod manufacturers. In other industries little is available to the public. In the important electronics industry, for example, many companies have developed their own standards for such things as printed-circuit-board assembly and harness forming, but what little has been published is too loose to be recommended. Sometimes an industry association has time data available. I know of one industry, glass bottles, where production speed information is updated and published by one equipment manufacturer. In some industries, although nothing on time standards has been published, it is possible to collect usable data by visits to industry meetings, by talking to equipment and supply salesmen, or talking to fellow estimators. Sometimes quid pro quo relationships can be established, in which you provide time standards you have in exchange for some you don't have. Research outside your company can often turn up better time standards to use.

For fabrication operations, where value is being added to material and machinery is being used, estimated operational times are almost inevitably in terms of machine time. Then when one operator is running one machine, the machine times are also labor times. When an operator is running more than one machine, the machine time must be factored by the number of machines being run by the operator to estimate labor costs. In that same situation the machine-hour time required for the operation may be the basis for developing other cost segments of the total estimate. For example, machine hours may be the base upon which manufacturing burden is applied. In assembly operations, estimated operational times are almost inevitably in terms of labor time.

Obviously the greater the labor portion of the total cost of sale, the more important good time standards become for adequate product-cost estimating. However, even when labor represents a smaller portion of the total cost of sale, the estimated labor cost is frequently the basis for applying manufacturing overhead. Or if machine hours are used, they may be the basis for manufacturing overheads. In either case, errors in the time estimates are exaggerated by the burden application, and the final total estimated costs become that much more in error—another reason for Estimating to urge senior management to have engineered time standards developed.

In conceptual estimating, details of operational routing and operational times are not applicable because the specifics of parts and assemblies are not available. Broad estimates of total machine and labor times must be guesstimated on the basis of past historical totals and/or on such broad product parameters as weight, capacity, tolerances, and the like. As in the case of material and routing, however, if some of the subassemblies will be the same as those used before, historical time costs of such subassemblies can be used by the estimator.

Step 6: Apply Labor and Manufacturing-Overhead Rates

In some companies Estimating's job is finished when step 5 has been completed, the material and labor and/or machine times needed for the job or product have been estimated, and the data go to another department, usually Accounting. There labor rates are applied to the estimated labor hours. Manufacturing overhead rates are applied to the estimated labor hour or (usually) dollars or to the estimated machine hours, depending on how manufacturing burden is applied.

This situation prevails in about 35 to 40 percent of the companies I am familiar with. I have never understood this practice. There is no reason why Estimating cannot be provided by Accounting with the applicable labor and manufacturing overhead rates, so that Estimating is responsible for their application and for development of the total manufacturing cost (step 7). This is the way it is done in about 60 to 65 percent of the companies polled at my seminars, and it has always struck me as the more sensible approach. Admittedly, it is more work for Estimating, but in these days of calculators and computers, the additional time burden is of little moment.

When it is done this way, only Estimating is responsible for calculating the total estimated manufacturing cost. Subdividing the job, as some companies do, divides the responsibility. When there are many different labor and overhead rates and/or great differences between them, the chance of error increases when the task is subdivided. Since Estimating is closest to the problem and knows more than Accounting can about the estimate, it is the natural choice to apply the labor and burden rates. Finally, when Accounting applies these rates, there is a natural tendency to oversimplify the labor and burden rates. As a result, too gross overall average rates are more likely to be used, which may overlook actual cost differences between the work and cost content of different jobs. Also, since Accounting is applying the rates, there is less likelihood that the rates will be questioned and improved if they are inadequate.

Labor rates are the dollar cost per labor hour. How these dollar rates are expressed and what they contain varies tremendously between companies. Smaller companies often use only a plantwide overall average cost per labor hour. Less frequently in smaller companies and more frequently in medium companies or divisions, an average hourly cost by department or work center is used. In larger companies or divisions, labor rates are likely to be applied by job position, skills, or crafts. How labor rates used are defined and constructed is a matter of balance. Certainly the fewer different labor rates there are, the easier they are to apply on the estimate. However, it is easy to oversimplify and turn out poor estimates. For example, if there are substantial differences in labor rates for different work in a given department, an overall average hourly labor costing rate for the department is likely to be completely inadequate. Certainly it will produce poor estimates where the work content by skills varies widely between different jobs. A worse example occurs when an overall plant average hourly labor cost is used although there are large differences in average hourly labor cost between departments or work centers.

Another difference in observable practice lies in how labor fringe benefit costs are handled and applied. The poorest procedure is to include fringe benefits in the overall manufacturing overhead rate because as these costs continue to increase it becomes more important to keep them segregated and handled as a percentage of labor costs. As such, they can be built into the individual labor rates applied either on the estimate or against the total estimated labor costs. I prefer to build fringe costs into the individual labor rates because it results in a total cost per labor hour, by job position, by subcenter, etc., according to how the labor dollar rate is developed.

When applied, manufacturing overhead rates load into the product-cost estimate the costs of owning and operating the plant (building costs, the plant manager, maintenance labor, equipment depreciation, production planning and control, operating supplies, etc.). We noted earlier that such burden rates must be by department or work center for adequate product-cost estimating. If you apply only one overall manufacturing burden rate, you assume the same burden for every center, which is rarely valid, and you are most likely to undercost some products and overcost others. It is not difficult for the controller and Cost Accounting to develop such departmental burden rates. Obviously it is more work than developing only one overall burden rate, but the yield for the extra work is great and the results in better product-cost estimating and thus better management decisions are important. The only mystery is why so many companies have not yet done it.

In all these matters of labor and overhead rates, Estimating must depend on the controller and on Cost Accounting under the controller for the data they have to use. Unfortunately the data and the breakdowns in those data are often not as good as they ought to be. The labor rates may be too broad averages—by departments, say, instead of by job position; or separate burden rates for costly subcenters within production departments may not have been developed for Estimating to use. If Estimating has this problem, it is their responsibility to expose the situation and push for better data and costing rates. This observation is easily made in a book, but in real life it can be hard for Estimating to exert such pressure. The controllership is a high and powerful position in most companies (and Accounting shares that power), and Estimating is often far below in the management hierarchy. Since an effective management is a structure of checks and balances, Accounting must service Estimating among all the others and Estimating must be the force to obtain the data needed.

Step 7: Calculate Total Manufacturing Cost

In this step all the cost factors determined in the previous steps are collected, burden rates are applied, and the total manufacturing cost predicted. It is the summing-up step. The work content of the step depends on the complexity of the product and the estimating detail being used. In very complex products, the bulk of the total calculation work involved in the estimate itself can lie in this step. For that reason some of the greatest gains achieved with the computer in Estimating have been realized in this area. In the years ahead many of the most attractive accuracy and time gains made will be in this step.

It is after step 7 in most companies (60 to 65 percent) that Estimating is finished with the estimate. The calculated manufacturing cost and estimate sheets then go to the Controller, Marketing, or wherever, where sales and administrative (S&A) burden rates are applied, profit markups are calculated, and the selling price is developed. From my seminar polling, I would estimate that in only 15 to 20 percent does Estimating apply S&A burden rates and standard profit markups to develop a desired selling price.

In only the minority of manufacturing situations and products are selling prices predicated absolutely on manufacturing costs. In such situations, to the calculated manufacturing cost are applied S&A burden rates and profit markups, to give the selling price that Marketing will use. You and I might think of this as the ideal or base situation. Such a price can properly be called a *standard selling price* because it is the price the company *should* get for the product if it is to cover all the product's costs and the burdens for manufacturing, selling, and administration and make what management has judged to be the profit needed for an adequate return on investment. But you and I know that this is rarely the situation in the real world. In many situations, while the estimated manufacturing cost is the starting point from which the ultimate selling price will be developed, a great many other matters affect the final selling price. For most companies, and with most products, in establishing the selling price actually to be charged senior management and Marketing must consider competitive pressures and accept less than the desired profit markup on certain products or models. Or, more happily, the company is able to sell at a greater than standard markup, at least for the time being, because of the existing competitive situation. In any case, the company should develop a standard selling price for reasons discussed below under step 9. We may never be able to charge that price, or perhaps we can charge more, but the standard should be developed.

Where and by whom should the standard selling price be calculated? It is an interesting phenomenon, this divorcement of Estimating from the development of the standard selling price. I think it often is the result of historical precedent within the company. When a company is small, the ownership often does not want the people doing estimating to know S&A burdens and markup percentages. As the company grows or is sold to a larger company, the precedent is followed. In very large companies, with the subdivisions of functions that necessarily occur, senior management often considers such work unsuitable for Estimating, and even in these cases it can be a matter of internal secrecy. I have never quite understood this thinking. Estimating deals with cost data, and cost data are confidential matters. Are S&A burden rates and markups so much more confidential? Any estimators worth their salt can find out S&A burdens within the company and knowing the manufacturing cost can work back from the selling price to determine what markups have been used. I see no reason why Estimating cannot and should not continue through step 9 and calculate the standard selling price right on the estimate. Then it is there and ready for senior management's use as specified below.

Step 8: Apply Selling and Administrative Burdens

Since total product costs must include the cost for Marketing and for the general administration of the company, S&A burden must be applied.

In a few companies manufacturing overhead and S&A burdens are all lumped together into one percentage application. Most of us would agree that this is poor practice. In effect it can result in capitalizing into inventory segments of the S&A costs which, to be prudent, are best expensed into the monthly profit and loss. This result is inevitable when all these overhead costs are lumped into one percentage because there is no way to segregate S&A costs and then exclude them from the value at which items are going into inventory or being taken from inventory. Also the procedure fails to isolate S&A dollars of cost clearly, and as a result these costs are less likely to receive management scrutiny and control.

The standard procedure is to expense S&A costs monthly on the profit and loss statement and build them into the total product cost by using a percentage or percentages applied on top of the estimated manufactured cost. Thus, the S&A costs are treated as costs on the monthly profit and loss statement and not included in the value at which units are put into, or taken out of, inventory.

More common is the use of one percentage to cover S&A costs, which are therefore isolated, expensed on the profit and loss statement each month, and not included in inventory valuations. It is a usable procedure but not the best possible. It does not give enough individual exposure to the two major areas of cost involved, selling individually on the one hand and administrative on the other. These two cost areas involve substantial chunks of money and should not be lumped into one total percentage. When isolated, they are more likely to receive attention and control action—not simply, of course, because two percentages are being used on the estimates. Better control is more likely to be exercised if the cost of each area is shown in the total product costs on an individual basis. Also, if it is possible to identify different selling costs for different products, separate percentages, one for selling and one for administration, should be used.

Probably the single most common error in applying selling burdens is the unjust application of selling costs among products that incur widely different selling expense. For example, the company applies one overhead selling burden rate on every product's estimated manufacturing costs. Or the total selling expenses are applied across the entire product line, as one percentage, and against each product's forecast sales volume, although some of those products require and receive a disproportionately larger share of the sales effort and cost. This is probably the most common single procedure followed. As a result, products needing the greater selling cost are being partly carried by the other products, and their true costs are being understated. The error is most obvious when a separate selling force is used for certain products within the company or division and the costs of that selling force are not charged to the products it sells but are lumped into one selling-burden percentage charged to all the products produced and sold. Neither of the above examples can exactly be called rare.

In such situations, the wise thing to do is to segregate selling costs between products (or at least families of products) and then allocate total selling costs logically to the various product lines. It is more work in cost reporting and recording and requires more logical thinking in selling-burden allocation, but if the money amounts involved are great enough it is the only way to determine realistic costs by product line. We are not concerned with price here. That is another matter. But to run the company wisely and make good management decisions, you need to know the total costs incurred by the various products. And that total cost includes selling costs.

Step 9: Apply Markups and Develop Standard Selling Price

The company is in business to make a fair return on the investment, and to do this the product has to be sold at a profit. The amount of profit is determined by the markups applied and ultimately, of course, by the final selling price. That price is the concern of senior and Marketing management. Although product pricing is not the subject of a book on estimating product cost, Estimating or some other management function, e.g., the controller's group, should develop a standard selling price.

As defined earlier, the standard selling price is that price at which, if the sales forecast is realized, the company will cover all the product's costs and all the manufacturing and S&A burdens assigned the product and will make the desired profit.

This standard selling price is frequently not achievable in real life. In competitive situations price concessions must be made. In some companies such concessions are rigorously defined and lie within narrow limits. In other companies, they are not as narrowly restricted and the final decision lies with Marketing. In any case, senior management, responsible for the profit and loss results, needs to control selling-price variances. To control, you need a standard, and clearly the standard to use is the standard selling price. Senior management can then track the difference between actual selling price and the desired or standard selling price. The point will be obvious to many readers. It is standard operating practice in their companies. But in a surprising number of companies such standard selling prices are not developed and senior management has no continuing control over the dollar variance being developed year-to-date by the prices actually being charged. The estimate sheet is the obvious document on which to record the standard selling price. The markup used for a given product or family of products should reflect the capital investment involved in making that product and its labor and material content. Thus, different products can have different standard markups. Many companies use one standard markup across all their product lines to develop the standard selling prices. This practice is satisfactory if all the product lines require substantially the same dollars of capital investment to produce *and* all the product lines have substantially the same ratios of material and labor costs, including purchased components.

To achieve a fair return on investment, the product requiring the greater capital investment should logically be given a higher markup and yield a higher profit. A product with a higher material cost (including purchased components) than another product can be unfairly overpriced with a too high standard selling price when the same markup is used for all products. Properly, a lower markup is used on material and purchased components than on the costs incurred within the company in making parts in its own plant and in assembling the product.

The development and application of markups reflects tremendous differences between industries and between companies within a given industry. Some companies do not even use standard markups to develop standard selling prices, using instead whatever markups will translate the total cost (including S&A) into a selling price that is competitive. Other companies do not have return-on-investment measures by product line, even though there are major differences between the capital investments required by the different products they make. Instead, they use, if any, only one overall return-on-investment measure. Still other companies use one overall markup for all product lines despite the fact that certain product lines have substantially higher ratios of material-cost content. These situations simply reflect the fact that much remains to be done in improving our management practices and the logic applied in the development of the standard product price.

Tremendous differences in practice between companies and divisions within companies can also be observed in how *both* S&A burdens and markups are applied to arrive at the suggested selling price. Consider one actual situation. The company is a metal-stamping manufacturer with two divisions, A and B. Manufacturing burden is 400 percent of direct labor in both divisions, and the company seeks a profit of 15 percent on all sales. Assume both divisions have a job on which:

Raw material costs	$40.00
Direct labor	5.00
Manufacturing burden (400%)	20.00
Total manufacturing cost	$65.00

Division A calculates the desired selling price:

Raw material	$40.00 + 15% profit ($40.00/0.85)	$47.06
Direct labor	5.00	
Manufacturing burden	20.00	
S&A = 35% of direct labor and overhead	8.75	
	$33.75 + 15% profit ($33.75/0.85)	39.71
Desired selling price		$86.77

Division B calculates the desired selling price:

Total manufacturing cost		$ 65.00
S&A 20%		
Profit 15%		
35%		
Desired selling price ($65.00/0.65)	$100.00	

Thus on an item with the same total manufacturing cost, the divisions develop two different suggested selling prices that are 13 percent apart. Division B is charging $11.25 more S&A and trying for $1.98 more profit than Division A—quite a competitive handicap. I would give all my stamping work to Division A.

In this chapter, the discussion has been in terms of conventional absorption costing, the kind used by the vast majority of our companies. In Chapter 4 this method was contrasted with direct costing. No matter which approach the company uses, there is still the need to assign segments of fixed costs (including S&A) to individual products to determine total costs. There is also still the need to determine standard selling prices by product. And in both instances the logic used must be agreed to by all the areas of management involved (top management, Marketing, Accounting, and Estimating) and fitted to the particular situation and product.

Life-Cycle Costing

The nine steps in estimating dealt with in this chapter are needed to determine the price the customer will be charged per unit for the product made, but in some product and industry situations today the manufacturer's estimating task is even broader. The customer also demands estimates of the cost to maintain the product over its projected life, commonly called *life-cycle costing*.

Today life-cycle costing is of greater and greater import in defense products. The government wants projections or estimates of the cost of spares and preventive maintenance and even breakdown maintenance work required over the life of the product. For commercial and defense aircraft engines and airframes, manufacturers can develop a reasonably good handle on the cost of spare parts. Design Engineering in such situations is able to specify the spare parts or spare kits it is advisable to carry. They have had experience with similar products in the past and know the specifics they have designed into the new model. The final factor in this cost element is governmental-agency requirements for spares; this is determined by the parts replacements called for by the agency's maintenance program requirements. For example, engines will be disassembled and inspected and certain parts replaced after so many hours of use. These same design determinations and government-agency requirements are also the chief determinant of the hours of preventive-maintenance labor. The really tough estimating job in life-cycle costing is making meaningful projections of breakdown-maintenance costs over the life of the product. Such breakdowns are the result not only of the manufacturer's design and quality but of operating conditions, customer operating practices, and the quality of the customer's preventive-maintenance work. Thus, on aircraft engines and airframes reasonably sound estimates should be possible for spare parts and preventive-maintenance hours over the product's projected life, but I, for one, would question the realism of estimates of breakdown costs or the practicality of recourse action against the manufacturer over the entire life of the product in question.

Life-cycle cost estimating is even more nebulous in defense electronics and avionic products. Spares and preventive-maintenance hours, like those for engines and airframes, are often predicated on the design and existing government-agency maintenance requirements. For other avionic products and many defense electronic products, it is basically the manufacturer who specifies (or at least strongly recommends) the spares and the preventive-maintenance action for the life of the product. Life-cycle breakdown cost estimates, if made at all, can be of doubtful value.

Using the Learning Curve in Estimating

The most important ingredients in learning curve performance are vision and leadership. Continued improvement is a chain of influences which start at the conviction that progress is possible, continues with the creation of an environment and support of work which promote it, and results in a flexibility and willingness to change established practices for more efficient ones as they continually evolve. Furthering this chain is part of the practice of management. Consequently the learning curve can be regarded as a primary tool of management.

W. F. Hirschmann†

The learning curve (LC) is a tool or technique that has found very important use in four areas of management:

1. Cost estimating
2. Production planning
3. Setting cost objectives and cost control
4. Vendor negotiations

† "Profit from the Learning Curve," *Harvard Business Review*, January 2, 1964. © 1964 by the President and Fellows of Harvard College. All rights reserved.

The greatest single use historically has been in cost estimating, the area of interest and application in this book. Everyone in management should become familiar with the learning curve, and everyone involved or interested in product-cost estimating should know how it is used in estimating. All practicing estimators should be conversant with it and alert to its possible application in their own estimating situations.

Use of the learning curve is standard operating procedure in certain industries. In the airframes industry, for example, it is used by every company. Elsewhere although it is equally applicable, it suffers from an extreme applications lag. It many companies where it has real applicability it is unknown.

Readers of this book will fall into three main groups: (1) those who are experienced in the learning curve's application in the estimating area, (2) those who use it rarely, and (3) those who have never used it nor seen it used in estimating. This tool is so new to so many practicing estimators that it is best to treat the subject in a step-by-step manner, working up from the basic concept to its actual application in product-cost estimating.

Since experience is helpful in developing an understanding of how the learning curve works, a series of case problems with solutions is provided at the end of this chapter. Readers who are not practiced in the use of the learning curve are strongly urged to do these problems as they are mentioned in the course of the chapter. All you need is your interest and a fair supply of 2- by 3-cycle log-log graph paper. Readers experienced in use of the learning curve in estimating are urged to try at least Problems 5 and 6.

History of the Learning Curve

In the early thirties in the United States there was great interest in private flying. Although these were terrible depression years, when good people who wanted to work could not find jobs, many Americans believed that just as they had (or hoped to have) a family car, the day would come when there would be a Henry Ford of the family airplane.

In Buffalo the Curtiss-Wright corporation had an airframe plant, managed by T. P. Wright, an aeronautical engineer. (The airframe is the plane without the engine. Today, it would be the plane without the engine(s) and the avionics but in those days, the word "avionics" had not yet even been coined.) Wright asked what would happen to the cost of the airframes he was making if he were to make more and more of them. In doing so, he did what you and I would probably do: he studied his past actual costs and what had happened to them as the quantity produced increased. And he detected a pattern. This was important, because a pattern would provide a rationale or basis to use in estimating the cost of future quantities.

Every time the quantity produced doubled, the cumulative average cost at the doubled quantity bore a fixed and lower relationship to the cumulative average cost at the previously undoubled quantity. Notice the critical word "doubled." The learning curve always applied to *doubled* quantities.

He published his findings in the February 1936 issue of the *Journal of Aeronautical Science*. His article, entitled "Factors Affecting the Cost of Airplanes," is considered the seminal work on the learning curve in manufacturing although observations on the phenomenon had apparently been made earlier:

• In *Planning Production Costs* E. B. Cochran quotes an unpublished manuscript by G. W. Carr, a California management consultant, which mentions use of the learning curve by Boeing in 1930 in negotiating the cost of the P12 airframe with the Army Air Force and the F4B airframe with the Naval Bureau of Aeronautics.

• Colonel Leslie McDill, commanding officer at McCoon Field, Dayton, Ohio, in 1925, reportedly identified the learning-curve relationship between quantity and manufacturing cost.

But Wright published the first article and is considered the father of the learning curve, which he named. (Since then it has also been called improvement curve, time-reduction curve, experience curve, manufacturing-time-forecasting or MTF curve, and manufacturing progress function.)

Wright's article caused no great furor in American manufacturing or management. It was just another article in a not too widely read technical journal. But suddenly the war started. Roosevelt announced in 1940 that the country intended to build 50,000 aircraft, a tremendous quantity. As a result, the airframes industry was expanding like mad. Lockheed was going into Georgia. Boeing was expanding for the first time out of the Pacific Northwest into Kansas. Ford was converting River Rouge. More airframes plants were being built or expanded in Southern California. The industry was under tremendous pressure from the government, who wanted to know, "How many planes can you deliver next month? The month after?" As a result, the airframe companies were desperate for a predictive tool. They adopted Wright's learning curve and used it with great authority and effect. They put a lot of talent to work on it and discovered refinements in it. Today, the learning curve is standard operating procedure in the airframes industry.

During and after World War II, the Air Force (not yet a separate service) collected cost data from the airframes manufacturers as they made more and more planes. This was an unrivaled opportunity in the history of the Industrial Revolution. Here were the cost data for many different types of a product (airframes) from many different companies and locations. (It is hoped that we never have another opportunity like this because it would probably take a war to collect such data.) The Air Materiel Command analyzed cost performances on 118 different models of airframes made in World War II. Subsequently, the Air Force contracted with the Stanford Research Institute (SRI) to make further analysis of the data. The SRI study encompassed 45 different cost performances. Out of these studies came these basic and important findings on the learning curve:

Different learning curves were developed for different airframes.

When the same airframe was made by different companies, the companies had different learning curves.

When the same airframe was made by the same manufacturer at different plants, those plants had different learning curves.

These variations in the learning curves have been supported by subsequent studies in the airframes industry made since World War II.

What do these percentages mean? Take a specific instance, the first one listed in Table 6-1, the B17, Douglass Aircraft at its Long Beach, California, plant: every time the quantity produced doubled, the new cost at the doubled quantity was 77.4 percent of the cost at the previous undoubled quantity. Thus, the learning curve is expressed as the complement of the cost-reduction percentage achieved between doubled

TABLE 6-1
Typical Airframe Learning Curves in World War II

Type	Manufacturer	Plant	Percent
B17	Douglass	Long Beach	77.4
	Convair	Fort Worth	76.4
	Lockheed	Burbank	65.3
B24	Douglass	Tulsa	75.0
	North American	Dallas	75.0
	Ford	Willow Run	70.0
B29	Boeing	Renton, first 400	80.5
		Last 400	79.0
		Wichita, first 900	71.8
		Last 800	69.5
AT9	North American	Dallas	98.0
P47N	Republic	Farmingdale	89.0

quantities. Obviously, for a given manufacturer, a 75 percent learning curve is preferable to an 80 percent learning curve.

Percentages are attractive but dangerous things. You cannot do a great deal with them. You cannot even compare them unless they are all calculated from the same base. Obviously those listed in Table 6-1 cannot have been calculated that way. For example, suppose that Douglass started out with excellent engineering, very good bug-free tooling, intensive operator training, and low initial costs. And suppose that Lockheed had great deficiencies in all those areas and high initial costs. In that hypothetical scenario Lockheed's 65.3 percent learning curve might not have resulted in lower or as low dollar costs as Douglass's 77.4 percent learning curve.

The average learning curve in the airframes industry in World War II was 80 percent, and you still hear the 80 percent learning curve referred to as the *airframes learning curve*. Of course, no airframes manufacturer necessarily uses this specific learning curve. Instead, each manufacturer develops knowledge of its own specific learning curves.

The Lesson from Learning-Curve History

We must never blindly adopt another company's learning curves. We must develop knowledge of our own learning performance in our own company. At a London seminar on estimating, an English estimator told me, "My company is doing major subcontracting work for your Lockheed Corporation. In developing our bids we used their learning curves, and we are in trouble." I asked, "You're not making them?" He replied, "That is correct." It can be dangerous to assume that you will be able to duplicate someone else's learning curves. Later in the chapter we show how to determine your own company's learning-curve performance.

Reasons for Learning in Manufacturing

There are two major categories of reasons for learning in manufacturing, operator improvements and (more important) management improvements.

OPERATOR IMPROVEMENTS

Here are some typical reasons for operator improvements:

• As the operator performs a given operation more and more often, it takes less study or thinking time. In many cases, the work steps become automatic or at least semiautomatic and are done without deliberate thought. Do you remember buttoning your shirt or blouse this morning? You did it automatically. But have you ever taught a very young child to put a button through a buttonhole? Sheer agony!

• As operators repeat the work on more and more pieces, they develop more efficient motions and more efficient personal methods, particularly if they can be motivated to do so. I had this impressed on me in my very first job in industry. I was working as a time-study engineer at a Bendix plant in Philadelphia, making aircraft starters and instruments. The job involved going from machine to machine throughout the plant, setting incentive standards right off the time-study board, but occasionally I had the time to do it right. I'd have the workplace layout ideally set up, the tool angles, sequence of cuts, and machine speeds and feeds done the best I could. And when I was new at the business, I'd set the rate and walk away believing that the operator would never beat that rate. I soon learned how wrong I was. That operator would think of things I never thought of.

Every time-study engineer and industrial engineer experienced in work measurement has seen this happen. In your company, if you think you are doing an operation the best that it can be done, you are kidding yourself. Someone will come along and show you how to do it better. There are no limits. Certainly, we have limits in the time available to determine and install improvements. Or perhaps the money benefits on the quantity involved may not warrant the effort and investment required. But the operation can be improved and should be if the quantities involved are great enough to pay for the improvement work involved.

MANAGEMENT IMPROVEMENTS

Typically as a company makes more and more of a given product:

1. Better tooling and methods will be developed and used.
2. More productive equipment will be designed and/or bought and used to make the product.
3. Increases in speeds and/or feeds will be achieved that reduce the product's work content.
4. Design bugs will be detected and corrected.
5. Engineering changes will decrease in number, or at least in seriousness.
6. Because of the money involved as product sales increase, Design Engineering will be prompted to achieve still better design to reduce material and/or labor costs.

7. The early material problems that always seem to occur will be overcome.
8. As the product involves more and more money, management will be more strongly prompted to better production planning and more and better management controls.
9. Rejects and rework will tend to diminish.

The result of all these reasons for both categories (operator and management) is that as the quantity produced increases, the cost per unit drops. The reasons are that each unit will entail:

1. Less labor
2. Less material
3. More units produced from the same equipment because of
   a. Fewer delays
   b. Less lost time
4. Less work in process

Of course, as a manager from Boeing said, "You have to work at it." The learning curve does not necessarily develop automatically although I know of some real situations in which learning curves were developed unconsciously. But companies effectively using the learning curve today use it intensively to establish successive cost goals as the quantities produced increase. The reader is referred to the quotation that led off this chapter. To paraphrase H. R. Kroeker and R. Peterson of Ohio State, declining learning curves indicate learning. Horizontal learning curves indicate no learning. Rising learning curves indicate suicide.

A Learning-Curve Example

The learning curve is the formulation of the following phenomenon. As the quantity produced of a given item doubles, the cost of that item decreases at a fixed rate.

The two key words in that statement are "doubles" and "rate." As the quantity produced doubles, the absolute amount of cost decrease will be successively smaller but the *rate of decrease will remain fixed.* The rate of decrease applies to *doubled* quantities.

The basic learning-curve formula is

$$y = ax^b$$

where y = cost of unit x
a = cost of first unit

x = any given unit
b = slope constant of particular LC being used

$$\text{Slope constant } b = \frac{\log \text{LC}}{\log 2}$$

Consider the following example:

No. of units	Worker-hours for unit	Reduction in worker-hours	Reduction, %
1	1000		
2	900	100	10
4	810	90	10
8	729	81	10
16	656	73	10

Clearly as the quantities in this example double, there is a 10 percent reduction in cost. Thus, a 90 percent learning curve is developing, because the learning-curve percentage is the complement of the percent reduction occurring as quantities double. Mathematically

$$\text{Slope constant } b = \frac{\log 0.90}{\log 2} = \frac{-0.04576}{0.30103} = -0.152$$

Thus

$$y = ax^b = 1000(x)^{-0.152}$$

and for, say, the sixteenth unit

$$y = 1000(16)^{-0.152} = 1000(0.656) = 656$$

Two things are happening in this developing pattern: (1) The *rate* of change is *constant,* and with the learning curve the direction of the change is always downward if learning (and cost reduction) is occurring. Therefore the b exponent is always negative. (2) The absolute hours of reduction are decreasing. You can readily visualize that at 32 units, 64 units, 128 units, and so on, the reduction in hours will become successively smaller. Nevertheless the hours of reduction are always positive, though smaller and smaller. In mathematical terms, we have an asymptotic curve. In studying this pattern, you might guess that at extremely large quantities the worker hours per unit would be in the decimals. This is not the case because of the *doubled* quantities involved. In fact, extending this particular learning-curve situation, we get

Units	1000	100,000	1,000,000
Hours	350	174	122

The effect of repeatedly doubling quantities is well illustrated by the old fable of the Persian king who wanted to reward his faithful vizier. He called him in and said, "For your years of great service, half the kingdom is yours. What do you want?" They were always giving away half the kingdom in these stories. And the wise old man took the chessboard and said, "All I want is a grain of wheat in the first square, two grains in the second, four in the third, until you have so used all the squares." And, of course, there was not that much wheat in the kingdom. It takes only 20 doublings of 1 to exceed a million; 32 doublings of 1 exceeds 4¼ billion, and there are 64 squares on a chessboard.

The pattern this cost experience is taking can be visualized by plotting the data on a piece of arithmetic graph paper, scaling units or pieces along the x or horizontal axis and hours of cost along the y or vertical axis. This arithmetic graph paper is the kind we generally use, on which equal linear distances along a given axis have equal values. In arithmetic graphing, it is considered the best practice, wherever possible, to start the x and the y scaling from zero. The cost data we are looking at are plotted in Figure 6-1a.

The data plotted in Figure 6-1a fall along a downward-curving, ever-shallowing line, but because it is an asymptotic curve, the line will never cross the x or horizontal axis. This is the appearance of a learning curve plotted on arithmetic graph paper.

Suppose we want to use this pattern to estimate what unit 300 will cost us. Right away we have some problems. Aside from the problem of needing a very big piece of graph paper, we have the more serious problem of trying to extrapolate curvilinear data by manually drawing a line. This is only for the foolhardy. Instead we can take advantage of the *constant rate of change* that is occurring. Whether we are doing work in biology, physics, chemistry, or business management, when our data have this characteristic, we can use another kind of graph paper and our data will fall right along a straight line. Extrapolating a straight line is easy; all we need is a good straightedge and a sharp pencil.

For this reason the kind of graph paper used in learning-curve work is log-log paper. When we graph this specific cost experience and 90 percent learning curve on such graph paper, the plot will fall along a straight line (Figure 6-1b).

Log-log graph paper is so vital to learning-curve use that it is important to specify its characteristics. In graphing the learning curve, units or pieces are always scaled along the x or horizontal axis. Hours or

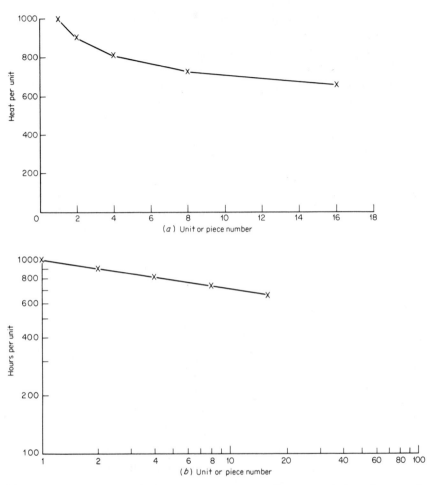

Figure 6-1 A 90 percent learning curve on (*a*) arithmetic and (*b*) 1- by 2-cycle log-log graph paper.

dollars of cost are always scaled along the *y* or vertical axis. The characteristics of log-log paper are as follows:

1. Along a given axis, equal successive distances have doubled values. In Figure 6-1*b*, on the *x* axis, the distance between 1 and 2 is the same as that between 2 and 4, 4 and 8, and so on. The reader may recognize that this corresponds to a given scale on a slide rule.
2. Unlike arithmetic graph paper, the scaling on log-log paper along a given axis is *never* done from zero. Instead you always start the scale from some power of 10, and if there is more than one cycle along

that axis, the next cycle is started with the next higher power of 10. In Figure 6-1*b* the *x*-axis scale starts with $10° = 1$, the next cycle starts with the next higher power of 10, or $10^1 = 10$. The next cycle, if there were one, would start with $10^2 = 100$.

3. Log-log paper is identified and ordered by the number of cycles on the *y* axis versus the number of cycles on the *x* axis. Since Figure 6-1*b* has 1 cycle on the *y* and 2 on the *x*, it is called 1- by 2-cycle log-log paper. In this chapter, we shall be using 2- by 3-cycle log-log paper.

It may well be some time since many readers last used log-log paper. Because it is integral to learning-curve work and because it is easy to make plotting errors with logarithmic scale reading, readers are urged to do Problem 1 at this point. It will not take long and may provide a cautionary illustration.

Identifying the Learning-Curve Slope

In any given learning-curve analysis, the rate of improvement, or slope of the line, on log-log paper is expressed as a percentage, which gives its name to that specific learning curve. The percentage is easily calculated by dividing the *y* value (hours or money) of the line at a given unit or piece quantity by the *y* value at one-half that quantity and multiplying by 100. Thus in the example given in the section above, at unit 2, 900 hours were used, at unit 1, 1000 hours were used, and $(900/1000)(100) = 90$ percent. Or, at unit 16, 656 hours were used, at unit 8, 729 hours were used, and $(656/729)(100) = 90$ percent. Thus the learning curve represented by these data is called a 90 percent curve.

Another way to calculate the learning-curve percentage if you know the *b* value is by the formula $LC\% = 2^b$. Thus in this example

$$LC\% = 2^{-0.152} = 0.90 = 90\%$$

Determining the Learning Curve with Nondoubled Quantities

When data for doubled quantities of units or pieces are not available, it is still possible to determine the learning curve being developed. Simply plot on log-log paper the cost data of the nondoubled quantities you have. Then draw a line through those plotted points and read off the *y* or cost values for doubled quantities.

For example, below are the cost data for a product at successively higher production quantities that are not doubled:

Unit no.	10	30	40	70	110
Hours	200	127	113	89	74

These data are plotted on 2- by 3-cycle log-log paper in Figure 6-2 and fall along a learning-curve pattern. What learning curve is being generated? We know that unit 10 cost 200 hours to produce. If we read off the cost indicated by the line for double 10 units, or for unit 20, the y or hours of cost value is 150 hours. (See the circled point in Figure 6-2.) Then, dividing 150 by 200 yields a quotient of 0.75, which multiplied by 100 yields a learning curve of 75 percent. In algebraic terms

$$LC \% = \frac{y_{2x}}{y_x} \times 100$$

Practice by doing Problem 2, where the learning curve being generated must be determined with nondoubled quantities.

Predicting the Cost of Future Units with the Learning Curve

Once we have identified the learning-curve pattern being generated, we can use it to predict the cost of still unmade quantities. We can

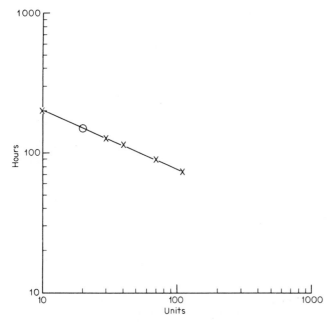

Figure 6-2 Plot of nondoubled quantities.

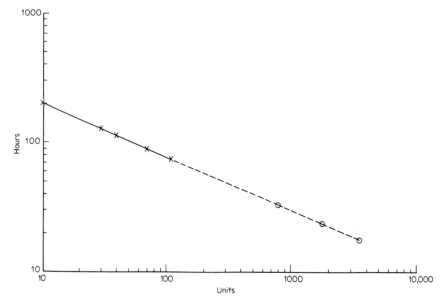

Figure 6-3 Predicting costs of future units.

illustrate this graphically for the example given in the previous section. We shall use the cost pattern being generated to predict (if this improvement continues) what the cost will be for units 800, 1800, and 3500. We read off Figure 6-3

Unit no.	800	1800	3500
Hours	32	23	18

But suppose this 75 percent learning curve is not continued and only an 85 percent learning curve is achieved after unit 110. The problem then becomes one of converting the 75 percent learning curve into an 85 percent learning curve beyond unit 110. This can readily be done. Twice 110 units is 220 units, and if unit 110 cost 74 hours, at an 85 percent learning curve unit 220 will cost 85 percent of unit 110 or 62.9 hours (Figure 6-4).

If we extrapolate the new and unfortunately shallower 85 percent learning curve to predict the cost of future units, we have:

Unit no.	800	1800	3500
Hours	46	38	33

Readers who think about this graphic procedure and particularly those who solve Problem 3 will realize that the thickness of the pencil used and how the line is drawn through the plotted points has a great effect on the values found for the cost of future units. Many people working with learning curves use learning-curve tables that are available for every learning curve from 51 to 99 percent. The procedure is to plot the cost data and eyeball a line through the plotted points, extending the line back to unit 1. This value is the a value in the formula $y = ab^x$.

By using triangles or reading off the costs at doubled quantities, as above, one finds the learning-curve percentage. Then the table for that curve is consulted. The table consists of a series of precalculated decimal factors for successively higher number of units, or x values. The decimal factor for a given unit number is applied against the a value, or the line's cost of unit 1, to calculate the cost of that given unit number.

Such tables have been a great boon over the years, but they are falling into disuse now that inexpensive hand calculators capable of using logarithms are available.

PREDICTING WITH A HAND CALCULATOR

The cost data are plotted and a line eyeballed through the plotted points. The learning curve being generated is calculated by reading off

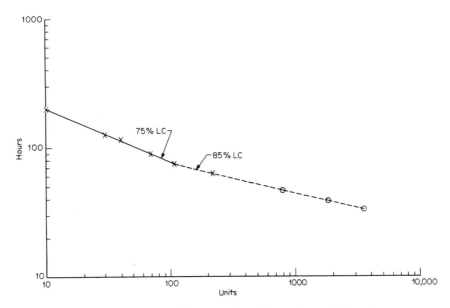

Figure 6-4 Predicting costs of future units with a changed learning curve.

the y scale the cost at doubled quantities. Then, as explained above, the line can be extrapolated backward on the graph paper to unit 1 to read from the plot the line's a value, i.e., the cost at unit 1 indicated by the line. A better way is to calculate the a value. When you know the learning curve involved, you know the slope constant to use. For a given point on the line drawn and a known slope constant, the calculated cost of unit 1, or "a," can be determined with the formula

$$a = \frac{Y_x}{\text{antilog } [(\log x)(\text{slope constant})]}$$

Remember

$$\text{Slope constant } b = \frac{\log \text{ of learning curve}}{\log 2}$$

Example On a 90 percent learning curve the cost of unit 16 is 328 hours. What is the calculated cost of unit 1?

$$a = \frac{328}{\text{antilog } [(\log 16)(-0.152)]}$$
$$= \frac{328}{\text{antilog } [(1.20412)(-0.152)]}$$
$$= \frac{328}{\text{antilog } (-0.18303)}$$
$$= \frac{328}{0.65611} = 500$$

Knowing a, the calculated cost of unit 1, and the slope constant, we can calculate the predicted cost for any future quantity with our hand calculator. The logarithm of the given unit quantity is multiplied by the slope constant for the learning curve involved. The antilog of this product is multiplied by the a value, and the product is the y value or cost for that given quantity.

Example For a given 90 percent learning curve the cost of unit 1 is 500 hours. Calculate the cost of units 450 and 950.

The basic formula is $y = ax^b$, or

$$y = a \text{ antilog } [(\log x)(\text{slope constant})]$$

For unit 450

$$y = 500 \text{ antilog } [(\log 450)(-0.152)]$$
$$= 500 \text{ antilog } [(2.65321)(-0.152)]$$
$$= 500 \text{ antilog } (-0.40329)$$
$$= 500(0.39510)$$
$$= 197.55$$

For unit 950

$$y = 500 \text{ antilog } [(\log 950)(-0.152)]$$
$$= 500 \text{ antilog } [2.97772(-0.152)]$$
$$= 500 \text{ antilog } (-0.45261)$$
$$= 176.34$$

It may look complicated, but with a little practice it is easier to use than the table for the 90 percent learning curve. Many users of the learning curve are now at the next higher stage of facility in fitting learning curves, as we shall see in the next section. First, however, the reader is urged to do Problem 3, dealing with matters covered up to this point in the chapter.

Least-Squares Fitting of the Learning Curve

So far, for illustrative purposes, we have worked with learning-curve data that fell nicely along a straight-line pattern when plotted on log-log paper. In real life it doesn't happen this way. When you plot real data for the learning curve, you always develop a scatter of the plotted points about (you hope) a line of central tendency. Once again we are faced with the problem we had in Chapter 3: Which line best describes the data plotted? What is the best description of central tendency to use? Again, we can eyeball a line, but you, I, and someone else could see three different lines, and still have the question: Who is correct? A better way is to use the method of least squares to describe the cost per unit relationship that is occurring.

Not every student of the learning curve advocates the use of least squares in learning-curve work. Paul McDonald in his excellent work *Improvement Curves* (see the Bibliography) holds that it "negates the advantage of the unit curve" and that it substitutes "mathematics for judgment." These points are well made if we use least squares only mechanically. But we still face the need of supporting and quantifying our human evaluations where possible. It is a matter of judgement, but faced with fitting a line on a scatter diagram, I advocate and use least squares. Perhaps from there further application of human judgment may be indicated in a given situation.

In this section we shall be doing least squares manually. This is the area of the next higher state of facility in fitting learning curves mentioned at the end of the previous section. People working with learning curves today can use such calculators as the Hewlett-Packard 67 or the Texas Instrument 59, which offer library programs in which the least-squares calculations are programmed into the machine (with the Hewlett-Packard it is called the *power curve*). Other learning-curve users have terminals on-line to the central computer or their own micro- or minicomputers programmed to calculate the line of least squares to the learning curve. That is, more and more of us have computational capability beyond our fondest dreams of the past, but since we need to understand what the machines are doing, we shall do it manually here.

Consider the following cost data:

Unit no.	2	5	9	15	35
Hours	440	350	325	300	160

This cost performance is plotted in Figure 6-5, and a typical scatter diagram is the result.

To fit the line of mathematical best fit, we need first to calculate the a and b values for the basic $y = ax^b$ formula for our specific data. Algebraical manipulation of this formula, where $N = 5$ occurs, gives us

$$\log a = \frac{\Sigma(\log x)^2(\Sigma \log y) - (\Sigma \log x)[\Sigma(\log x \log y)]}{N[\Sigma(\log x)^2] - (\Sigma \log x)^2}$$

$$\log b = \frac{N[\Sigma(\log x \log y)] - (\Sigma \log x)(\Sigma \log y)}{N[\Sigma(\log x)^2] - (\Sigma \log x)^2}$$

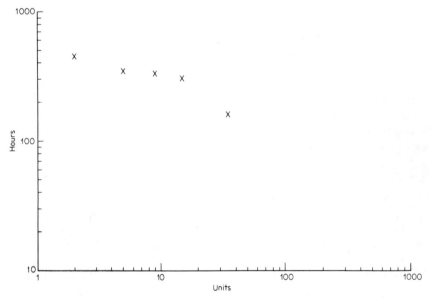

Figure 6-5 Costs at successively greater numbers of units.

x (unit no.)	y (hours for unit)	log x	log y	log x log y	(log x)²	(log y)²
2	440	0.30103	2.64345	0.79576	0.09062	6.98784
5	350	0.69897	2.54407	1.77823	0.48856	6.47228
9	325	0.95424	2.51188	2.39695	0.91058	6.30956
15	300	1.17609	2.47712	2.91332	1.38319	6.13613
35	160	1.54407	2.20412	3.40332	2.38415	4.85814
		4.67440	12.38064	11.28758	5.25710	30.76395

$$\log a = \frac{5.25710(12.38064) - 4.67440(11.28758)}{5(5.25710) - (4.67440)^2}$$

$$= \frac{65.08626 - 52.76266}{26.28550 - 21.85002}$$

$$= \frac{12.32360}{4.43548}$$

$$= 2.77841$$

$$a = 600$$

$$\log b = \frac{5(11.28758) - 4.67440(12.38064)}{5(5.25710) - (4.67440)^2}$$

$$= \frac{56.43790 - 57.87206}{26.28550 - 21.85002}$$

$$= \frac{-1.43416}{4.43548}$$

$$= -0.32334$$

Thus the formula is

$$y = 600(x)^{-0.32334}$$

The least-squares line can now be drawn by calculating the y value for one or two additional points.

For 2 units

$$y = 600 \text{ antilog } [(\log 2)(-0.32334)]$$
$$= 600(0.79922) = 480$$

Thus, the learning curve is

$$\frac{y_2}{y_1} \times 100 = \frac{480}{600} \times 100 = 80\%$$

or

$$LC\% = 2^b = 2^{-0.32334} = 0.80 = 80\%$$

For 100 units

$$y = 600 \text{ antilog } [(\log 100)(-0.32334)] = 135$$

While we need only the y value for two x points to draw the least-squares line, it is good practice to have the check of three calculated points falling along a straight line, and we now have them:

Unit no.	$1 = a$	2	100
Hours	600	480	135

The least-squares line for our cost data making the scatter diagram in Figure 6-5 can now be drawn. The y values for these three x points are shown as circled points in Figure 6-6. In this case there obviously is an excellent correlation for the learning-curve formula $y = ax^b$ applied to this given cost experience. The line is a good fit. How good a fit can be quantified by calculating the coefficient of determination r^2 and its square root, the coefficient of correlation r.

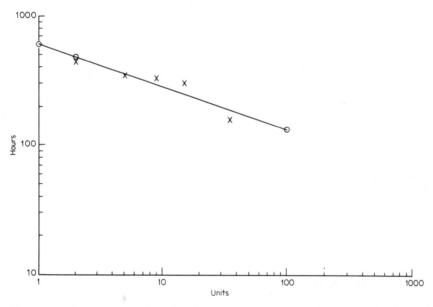

Figure 6-6 Least-squares line fitted to costs at successively greater numbers of units.

The formula is

$$r^2 = \frac{\left[\Sigma(\log x \log y) - \dfrac{(\Sigma \log x)(\Sigma \log y)}{N}\right]^2}{\left[\Sigma(\log x)^2 - \dfrac{(\Sigma \log x)^2}{N}\right]\left[\Sigma(\log y)^2 - \dfrac{\Sigma(\log y)^2}{N}\right]}$$

$$= \frac{[11.28758 - 4.67440(12.38064)/5]^2}{[5.25710 - (4.67440)^2/5][30.76395 - (12.38064)^2/5]}$$

$$= \frac{(-0.28683)^2}{0.88710(0.10790)}$$

$$= 0.85953$$

$$r = 0.92711 \quad \text{an excellent correlation}$$

All these laborious calculations can easily be made by the proper hand-held calculator like the HP-67 or TI-59. You simply enter, successively, the unit number and the cost for that unit, and the machine does all the work. Therefore, if you have any appreciable amount of learning-curve work to do, the more advanced and expensive calculator soon pays for itself. (And its cost is decreasing because of the learning curve.)

Great differences can be evolved in predicting the costs of future units when various eyeballed lines are used, as demonstrated in Problem 4. Do it now. The differences between the least-squares line and your eyeballed line may be startling.

The Two Types of Learning Curves

When Wright wrote the first paper on the learning curve, he formulated the *cumulative average learning curve*. Thus, he said, whenever the quantity produced doubled, the new cumulative average cost at the doubled quantity bore a lower and fixed relationship to the cumulative average cost at the previously undoubled quantity. When these cumulative averages were plotted on log-log paper, they fell along a declining straight-line pattern.

In subsequent use of the learning curve, particularly by the airframes industry in World War II, a further refinement was developed, the *unit-cost learning curve*. In the examples used thus far in this chapter and in the problems unit-cost data were used, namely, the cost at specific unit numbers.

For a given product, if the cumulative average costs most closely approximate a straight line on the log-log plot, the unit costs will then follow the pattern shown in Figure 6-7. Thus, as the cumulative averages fall along a declining straight-line pattern, the unit costs will flare

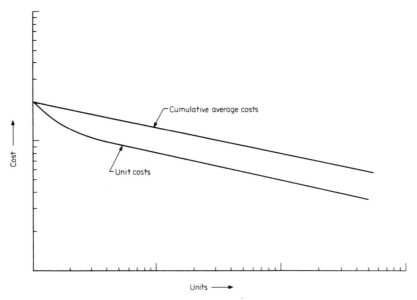

Figure 6-7 Relationship between unit and cumulative average costs with a cumulative average learning curve.

out and below the cumulative averages, until, usually between 30 and 50 units, their plot will be approximately parallel to the cumulative average plot. The higher the learning-curve percentage, the sooner the two lines begin to parallel.

Conversely, if for a given product the unit costs most closely approximate a straight line on the log-log plot, the cumulative average costs will follow the pattern shown in Figure 6-8. Thus, as the unit costs fall along a declining-straight-line pattern, the cumulative average costs will flare out and above the unit costs until, usually between 30 and 50 units, their plot will be approximately parallel to the unit-cost plot.

The fact that these two lines are parallel after the early quantities allows us to convert from unit cost to cumulative averages, or vice versa, as necessary. For example, in many estimating and in most purchasing vendor negotiation situations, cumulative averages are needed or simpler to use but our curves may be unit-cost curves. The conversion is easily done with conversion factors that vary with the learning curve being used (Table 6-2). For a given curve, to convert from cumulative average to unit costs, the cumulative average is multiplied by the decimal conversion factor. To convert from a unit cost to the cumulative average, the unit cost is divided by the decimal conversion factor.

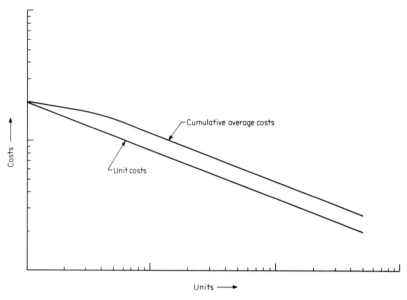

Figure 6-8 Relationship between unit and cumulative average costs with a unit-cost learning curve.

Thus with an 80 percent learning curve and a cumulative average cost at unit 100 of 200 hours, the unit cost of unit 100 would be (200 hours)(0.678) = 135.6 hours. Conversely, if with a 90 percent learning curve the unit cost at unit 100 was 150 hours, the cumulative average cost to unit 100 would be (150 hours)/0.848 = 176.9 hours.

Most learning-curve practitioners, but by no means all, use the unit-cost learning curve. It is preferred because it is the most sensitive. Thus, as you make more and more of a given product and plot your unit costs, if the line begins to shallow out you can promptly see that your

TABLE 6-2
Conversion Factors for Unit and Cumulative Costs

LC %	Factor	LC %	Factor	LC %	Factor	LC %	Factor
75	0.585	81	0.696	87	0.799	93	0.895
76	0.664	82	0.714	88	0.816	94	0.911
77	0.623	83	0.731	89	0.832	95	0.926
78	0.642	84	0.748	90	0.848	96	0.941
79	0.660	85	0.766	91	0.864	97	0.956
80	0.678	86	0.782	92	0.880	98	0.971

rate of cost improvement is declining. In contrast, cumulative averages always include past cost performance and any shallowing out of the learning curve tends to take longer to detect. Today, most companies that use the learning curve most effectively use unit-cost curves.

On occasion there still is controversy over which curve to use. Devotees of one sometimes tend to critize the users of the other. In a given product situation, after the early pieces, I like to plot *both* cumulative average and unit-cost data, fit lines, calculate which gives me the better correlation, and use that one. This can result in different learning-curve slopes, one for the cumulative-average line and one for the unit-cost line. Then your estimating judgment must be applied to choose the one you consider most applicable. If you are in doubt, use the curve with the best coefficient of correlation. In the months that follow you should be tracking subsequent actual cost performance as additional units are made and testing the fit of both lines. At any time, however, whichever line you select and are using, after approximately 50 units, the conversion factors can be used.

Continuing the Learning Curve

One of the great mental hurdles for the estimator and manager unfamiliar with the learning curve is the belief that the line must eventually flatten out. The only answer is, "Why does it *have* to?" Although there are factors and events that cause it to stop, if the economics are different, stopping need not occur. Here are two examples.

• One automobile company, when it starts its assembly lines at a new model year, gradually builds up a given line's speed over the first 4 weeks; then it stays there for the rest of the year. The leveling out is for physical plant reasons. Increasing the line speed further would require reducing the work content of assembly stations with the longest work cycle time. That would require more stations, and there are physical limits to the length of the existing building and assembly line.

• One jet-engine manufacturer had a program under which only 250 engines were to be made. A definite learning curve was developed for the first 200 engines, and it flattened out for the last 50. The generally accepted explanation within the company was that after the two-hundredth unit everyone knew that only 50 more were to be made and so further cost-reduction effort on that program ceased. That same company had other programs where the learning curve was continuing as production passed 1000 and 1500 units.

Where the learning curve does level off though production continues, probably the greatest single reason is management inertia. Management is just not intent enough on achieving continuing cost improvements. To accept that the learning curve *must* flatten out is to accept that we cannot improve our designs, tooling, methods, controls, and all the other factors that, when improved, cause the learning curve to persist. Management's attitude should be that there is no limit.

· One plant of the manufacturing division of a major communications company made an electrical-mechanical assembly and followed a 73.6 percent learning curve from unit 10,000 to unit 500,000, the total needed.

· In terms of a given industry, glass bottles are a good example. In this country, we make billions of glass bottles each year. In one or more consulting assignments in each 5-year period over the past 35 years I have seen measurable evolutionary cost improvements in such areas as bottles per machine cycle, bottle-forming speeds, Btus per ton of glass made, square feet of melting-tank surface per ton of molten glass pulled, melting-tank life, select and pack methods, etc. That is the learning curve at work over billions and billions of glass bottles made in the last three decades. It has never been measured, but it has occurred.

Setting Cost Objectives with the Learning Curve

The learning curve can be an effective way to establish cost objectives in making a given product. This can best be exemplified by the case of an electronics company which did not use the learning curve but had a big subcontract for a major office-equipment manufacturer. The subcontract was for a $250 printed-circuit board on which they were assemblying 200 components and 9 LED readouts. One industrial engineer working there kept a track of assembly hours on his own; the company was not using learning curves. *Unconsciously* the company was developing a definite 93.8 percent learning curve, out through 43,000 units. On such a big contract, management was making cost improvements and that was the result. Think about this case. When you multiply $250 by 43,000, you already have a $10,750,000 contract. If management had been using learning curves and setting cost objectives with the curve, they might have achieved a 90 percent curve or an 88 percent curve. In fact, an 85 percent curve on printed-circuit-board work is not uncommon, and I have heard of an 83 percent curve actually being achieved. Obviously the profit on a $10,750,000 contract with an 88 or 90 percent learning curve is substantially greater than with a 93.8 percent curve.

Setting cost objectives with the learning curve can be a subtle point to a management unused to this tool. As they get into a new product line, most managements work and push to reduce costs—the more money involved in the product, usually the greater the effort expended. But goals or standards are needed, and the learning curve can be an excellent goal-setting tool.

An example of this was given me at an estimating seminar. After the learning-curve discussion a participant came up and said that now he knew about learning curves he understood what was happening at his company, which was making a new amusement park ride and had orders to date for 12 rides. The company was not using the learning curve but quite properly were collecting their costs for each unit made. The first unit took 1200 hours to build. The fifth unit, the latest one made, took 700 hours. Thus the company was unconsciously tracking along a 79.3 percent learning curve, which may be quite good. But several questions are worth asking:

• How are they going to "ride herd" on this program to be sure that this trend is continued until the twelfth unit and further, as more orders are received? If they succeed, the twelfth unit will cost 522 hours, but if it shallows out to an 85 percent learning curve after the fifth unit, the twelfth unit will cost 570 hours. A 90 percent learning curve means that the twelfth unit will cost 613 hours.

• Is the 79.3 percent learning curve they are achieving from the first to the fifth unit the best achievable? Might a 78 or a 75 percent learning curve be possible? These are questions worth asking and goals worth striving for because a 78 percent learning curve would make the twelfth unit cost 492 hours instead of the 522 hours of the 79.3 percent learning curve. A 75 percent learning curve would make the twelfth unit cost 428 hours.

Using the Learning Curve for Cost Control and Cost Improvements

Related to setting cost objectives with the learning curve is its use for controlling manufacturing costs and achieving cost improvements as more and more units are made. Consider some management logic that you and I have encountered frequently. We are all too experienced to assume that unit 1 of a new product line will ever be made at standard cost. Instead, we speak of *start-up costs* and *learning*; but if we are good managers, we establish some future unit quantity at which we shall work to achieve standard cost.

Assume that our objective is to reach standard cost at unit 200. Being good managers, we track our actual costs as we make more and more of the product, and, with some scatter, we are headed in the right direction toward our goal. And we are pleased—as we should be. We continue to make more and more units and at the two-hundredth unit we are at standard cost. And we are pleased—as we should be. The situation is illustrated in Figure 6-9. We continue to make more and more of the products, and with some scatter, we continue at standard cost. We have no unfavorable variances. The illustration continues in Figure 6-10.

And we are pleased! Who can be unhappy with no unfavorable cost variances? But why did we suddenly stop learning at unit 200? This is a curious logic, harshly stated here, but it is all too common in real life. As we make more and more units, their costs should be tracked; when the control shows a shallowing out or, worse, a flattening, cost-improvement action is indicated.

Some Characteristic Learning Curves

Learning curves are highly individualistic, and each company must determine its own curve or set of curves by collecting and analyzing

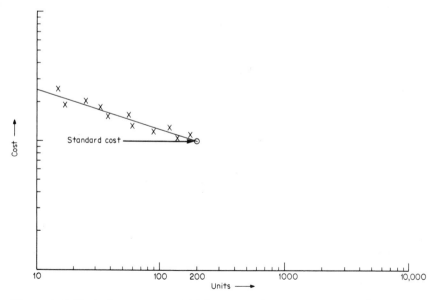

Figure 6-9 Typical management logic on cost improvements: product startup.

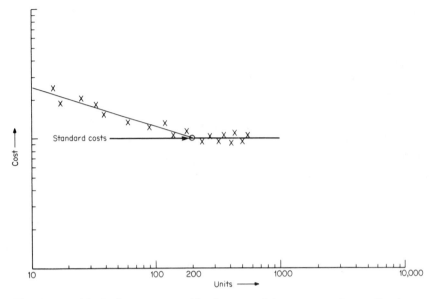

Figure 6-10 Typical management logic on cost improvements: continuing production.

actual cost performances on individual products. Your cost records must therefore be established so that costs are segregated by product and stated for given product quantity increments. Then these cost records are analyzed to determine costs for successively higher quantities of production to identify the learning path being generated.

The type of industry or operation has a major influence on the learning curve generated. For example, consider only machining and assembly. As the machining content of a product's cost is a higher proportion of total costs, the rate of learning declines and the learning curve has a higher percentage value. Conversely, as observed earlier, the greater the hand-assembly content of the total work needed to make the product, the lower the learning-curve percentage normally encountered. Writers and practitioners of the learning curve have used the learning curve within the following ranges in the indicated areas:

Machining	100–90%
Fabricating, machine	95–80%
Fabricating, hand	90–75%
Hand assembly	90–60%
Purchased material	100–80%

There obviously is no unanimity of finding or of application.

You must develop knowledge of your own learning curve(s) in your own plant. To accept someone else's as your own can be costly, even critical. As general guidelines, the following learning curves represent a rough estimate of learning-curve objectives:

Raw material	95%
Purchased parts	95%
Machining	90%
Fabricating, machine	90%
Fabricating, hand	85%
Assembly	80%
Direct labor (not classified)	87½%
Engineering	90%
Printed-circuit-board assembly	85%

Developing Your Own Learning Curves

LOT-MIDPOINT METHOD

For estimating purposes, the better learning curve to use is the unit-cost curve. To develop such a curve, it is necessary to develop data of cost for given higher unit numbers produced. Thus, for a successively higher series of unit numbers we need the cost to produce each of those units, but in many manufacturing situations the cost records do not enable us to determine the cost for each specific unit. Instead, batches or lots of units are produced and costs collected for each lot. Nevertheless there is a method of using lot-cost records to develop unit-cost data.

With one important exception the average cost per piece for a given lot can be applied to the median, or middle unit number, of the lot. Thus, all we need to calculate is the midpoint (median) of the particular lot in question and apply to it the average cost per piece for the lot. Then we treat our data exactly as we do in the unit-cost learning curve.

Consider an example:

Lot no.	Pieces in lot	Hours for lot	Cost average piece	Lot midpoint
1	8	800	100	4
2	20	1800	90	18
3	40	3200	80	48

Thus you would plot:

Unit no.	4	18	48
Hours avg. piece	100	90	80

Newcomers to the learning curve sometimes have trouble with lot midpoints. One way to overcome this mental hurdle is to think of attaching a serial-number tag to each unit made, starting with 1 on the first unit, 2 on the second unit, and so on. After the first lot in the above example you have used eight serial numbers. Halfway through the second lot, you will be applying serial 18, or 8 plus half of the second lot's 20 units. After finishing lot 2, you will have used 28 serial numbers and therefore the midpoint of lot 3 is 28 plus half of the third lot's 40 units, or 20 more serial numbers. Thus $28 + 20 = 48$ is the lot midpoint for lot 3.

The exception mentioned above occurs when the initial production lot is 10 pieces or more. Then we take not the midpoint of that first lot but the unit at one-third the way into the first lot. This exception applies only to the first lot, and if the first lot is less than 10 pieces, we take the midpoint as we do in all subsequent lots. For example:

Lot no.	No. in lot	Cumulative total	Lot midpoint
1	15	15	5
2	20	35	25
3	30	65	50
4	40	105	85
5	50	155	130
6	100	255	205
7	100	355	305

To summarize:

1. On the first lot
   a. If less than 10 units, use the midpoint
   b. If 10 or more units, use the unit at one-third into the lot.
2. On all subsequent lots use the midpoint.

All these rules of thumb have found practical application in many real-life situations. There is a difference between the rule-of-thumb lot midpoint and the mathematical or true lot midpoint, but in many applications the differences have been found to be insignificant. As you become more involved with the learning curve and use it more intensively, calculating the mathematical lot midpoint is indicated.

Estimating with the Learning Curve

By now it should be apparent how and why the learning curve is such a valuable estimating tool. It provides a rationale for predicting on the basis of past cost trends what future units should cost. The varieties and subtleties of such learning-curve use are limited only by the user's experience and vision. We list but a few examples:

Based upon product costs to date, what should the cost of future given units be?

Do learning curves actually achieved within the company vary markedly between different types of products? If so, why? On the products with higher learning-curve percentages can some techniques or management approaches successful on other products lower the percentages?

On the basis of in-house experience on similar product programs, how large a quantity must be produced before standard costs are achieved?

Are the costs of a major engineering change warranted in the light of foreseeable future production needed for a given product? Or (the obverse) how much additional production is needed to overcome the higher costs that will result from an engineering change? Or what improvement in the learning-curve shape should be realized to pay for the estimated costs of a major engineering change?

If the production of a product is interrupted, how much higher will the product's costs be when production is resumed?

If the learning curve of a given product shallows out, what will be the cost effect on units yet to be made?

To justify the additional costs of better equipment and/or tooling, what improvement in the learning curve and/or additional production at the present learning curve will be needed?

For a small order on a new product, custom-made to a customer's specifications, what is the cost estimate upon which to base a fair price?

A counterexample to this last use is the electronics plant of a well-known company. I visited a friend working there, who introduced me to a fellow industrial engineer, a student of the learning curve. We exchanged some learning-curve references and I commented that the company must be using his knowledge a great deal. His reply was, "No, we have a costing-pricing group here that will charge you as much for twelve as they will for two." This is lovely if you can get away with it, but it is very poor costing and pricing.

The uses for learning curves in estimating are varied, almost limitless. Certainly Estimating's use of the learning curve is bound only by the estimator's knowledge, interest, and energy.

Now do Problems 5 and 6. Even estimators who know a lot about the learning curve are urged to do Problem 6.

Further Study of the Learning Curve

There are too many subtleties, variations, and uses of the learning curve to cover here. Practicing estimators who are not using the learning curve in their own work are urged to experiment and investigate in their own milieus. Collect cost data for given product lines, analyze them, and see whether a learning-curve pattern can be detected. Use your wit and ingenuity in your present estimating situation to find practical uses for the tool. If your product has many parts, some of which you fabricate, and/or assemble, and/or test, the odds are superb that the learning curve is applicable.

Read more on the subject. There are many more aspects of the learning curve than we have reviewed in this chapter. For example,

• The discussion has dealt with straight-line learning curves. For many of us, in many actual situations, the data available only allow such a learning curve, but there are variants of the basic learning curve.

1. The *hump curve* may be a better description of the cost experience when plotted on log-log paper. When very good planning, tooling, and preparation have characterized preproduction, the costs over the first few units may decline along a flatter shape. After these early units, a more normal and steeper learning curve may be achieved. Thus the name "hump curve."
2. Cost trends over a long period may follow an S pattern on a log-log plot with three definite stages, namely, a relatively shallow slope at the start, a quite steeper slope following the first stage, and then an intermediate slope continuing indefinitely.
3. The Stanford Experience Curve applies β factors that vary inversely with the experience of the company in manufacturing the product.

• In many companies Estimating is involved with other management areas. Some estimators wear more than one hat. For example, the learning curve has found very effective use in vendor negotiations. You, as the estimator, may find opportunities for constructive suggestions for using the learning curve in areas other than estimating.

• By recording and analyzing your cost records for given products or families of products, determine your own company's learning curve or curves. Then try to determine the cost effect of such real-life occurrences as major engineering changes and gaps in the product's production. Such occurrences will cause blips, hiccups, or scallops in the learning curve. For example, if a product is not made for some months, when production is restarted, the odds are excellent that the unit costs will be higher than when the production was stopped. How much higher? In the experience of one practitioner this is a function of the prior production rate times the months of hiatus. Thus, if the last unit made was unit 100, the production rate was five per month, and the product was not made for 3 months, the cost of the first unit made after restart will be equal to the cost of the eighty-fifth piece. That formula may not apply to your operation, but it is worth investigating.

Case Problems in Learning-Curve Work

The reader will need a supply of 2- by 3-cycle log-log 8½- by 11-inch paper for these problems, obtainable from your Engineering department, local engineering supply dealer, a good stationer, or the bookstore of your local college or university.

In these problems, for the sake of clarity, costs are stated in terms of hours. In many situations of learning-curve use, money costs are the only ones available or the ones more readily available. With money costs, indexing to a base period is required to obtain comparable figures. It's an additional complexity that must be handled.

PROBLEM 1: LOG-LOG PLOTTING

The following 26 data points are to be plotted on a piece of 2- by 3-cycle log-log paper. Since units are scaled along the longer x axis, the 3-cycle axis must encompass units from the least (31) to the largest (3550). Thus the x-axis cycles will be 10, 100, and 1000. Since the smallest and largest numbers of hours to be scaled along the y axis (2 cycles) are 115 and 7000, the cycles on the y axis will be 100 and 1000.

Simply plot the 26 points. They will appear quite random as you do them, but when they are all plotted, if your errors are few, a pattern should be visible. A picture of the perfect plot is in the Answers section at the end of this chapter. Incidentally, one definition of an optimist is a person who does this kind of plotting in ink.

Units	Hours	Units	Hours
31	7000	580	1100
105	215	41	3150
3550	145	1850	1700
215	1450	145	1250
3000	1450	480	2300
295	1050	1100	120
780	1400	48	2000
55	1350	1150	1650
610	165	440	720
400	450	96	1070
36	4700	180	500
780	4800	450	255
2050	115	63	910

PROBLEM 2: DETERMINING THE LEARNING CURVE WITH NONDOUBLED QUANTITIES

The cost records available on a given product show the following cost performance:

Unit no.	2	5	15	50	120
Hours	150	112	78	53	40

What learning curve is being developed by this cost performance?

PROBLEM 3: PREDICTING THE COST OF FUTURE UNITS

Cost of early quantities on a product are as follows:

Unit no.	15	39	61
Hours	250	200	180

Plot these data on 2- by 3-cycle log-log paper with the x or unit axis cycles scaled 1, 10, and 100 and the y or cost axis cycles scaled 10 and 100.

(a) What learning curve is being generated by this cost performance?

(b) What is the b value or slope constant?

(c) Extrapolate your plot back to unit 1 or a in the learning-curve formula. What is your reading? Now calculate the cost of unit 1 with the formula

$$a = \frac{y_x}{\text{antilog } [(\log x)(\text{slope constant})]}$$

(d) State the formula for this particular learning curve.

Next, make a new plot of this data on 2- by 3-cycle log-log paper with the x or unit axis cycles scaled 10, 100, and 1000 and the y or cost axis cycles scaled 10 and 100, as before.

(e) Extrapolate your plot forward into the 1000 cycle. From your line read off the predicted costs for units 170, 620, 1400, and 3000. Now calculate these costs with the formula

$$y = a \text{ antilog } [(\log x)(\text{slope constant})]$$

Finally, suppose that after unit 61 only a 90 percent learning curve is achieved.

(f) Calculate the new slope constant.

(g) Draw the new line on your log-log plot. Read off the line's value for unit 1. Then calculate this value.

(h) State the formula for the new learning curve.

(i) Extrapolate graphically the new learning curve and read off the predicted cost for units 170, 620, 1400, and 3000.

PROBLEM 4: LEAST-SQUARES FITTING OF THE LEARNING CURVE

Plot the following cost experience:

Unit no.	2	5	7	20	35
Hours	400	350	250	230	180

By eye alone, using a straightedge, draw the line you think best fits the plotted data. What learning curve have you plotted? Extend your line out to unit 800. What would be your prediction of the cost of unit 800?

Now apply least squares and calculate

(a) a, the cost of unit 1

(b) The slope constant

(c) The learning-curve percentage; compare with your eyeballed line

(d) The predicted cost of unit 800; compare with your eyeballed line's prediction

(e) The coefficient of correlation r

PROBLEM 5: ESTIMATING WITH THE LEARNING CURVE

Your electronics company is bidding on a major subcontract for a blackbox which your company would assemble and test. You estimate that a total of 100 hours at standard will be needed for all the subassembly and final assembly and intermediate and final testing. The company has done similar work in the past, and what cost records you have available indicate a 90 percent learning-curve experience on such work. Finally, you (the estimator), Production management, and Industrial Engineering all agree that your company should achieve standard costs at unit 100. Thus, three things are determined: a 90 percent learning curve should be possible, standard cost should be 100 hours, and the standard should be achieved at unit 100.

The final contract quantity cannot be settled yet by the customer, who has requested bids for various possible quantity levels. As a result, to develop your bid, you need to estimate the total labor hours needed for **(a)** 25, **(b)** 50, and **(c)** 500 units.

PROBLEM 6: ESTIMATING WITH THE LEARNING CURVE

The cost data for the first seven consecutive production lots of a given product are given below. First calculate the rule-of-thumb lot midpoints to develop your unit cost data. Then on a single piece of 2- by 3-cycle log-log paper plot both the unit-cost data and the cumulative average cost data.

Lot no.	Pieces in lot	Total hours for lot	Lot midpoint	Hours avg. piece	Cum. pieces	Cum. hours	Cum. avg.
			Unit-cost data		Cumulative cost data		
1	15	15,000	_____	1,000	15	15,000	1,000
2	20	13,720	_____	686	35	28,720	821
3	30	17,490	_____	583	65	46,210	711
4	50	25,400	_____	508	115	71,610	623
5	50	22,900	_____	458	165	94,510	573
6	100	41,400	_____	414	265	135,910	513
7	100	37,900	_____	379	365	173,810	476

(a) What unit-cost learning curve is being generated?

An eighth lot consisting of an additional 635 pieces is going to be made, bringing the total produced up to 1000 units.

(b) Estimate the unit cost at unit 1000.

(c) What will the cumulative average cost be at unit 1000?

(d) Estimate the average cost per piece for the additional 635 units from unit 366 through unit 1000.

(e) After unit 365 if only a 90 percent learning curve is generated to unit 1000, what would be the additional cost, at $10 per hour, for the final 635 units, compared with the cost for those 635 units if the learning curve generated up to unit 365 continued out to unit 1000?

Answers to Learning-Curve Case Problems

PROBLEM 1

The plot of the 26 points should take the pattern shown in Figure 6-11. At most, 2 percent of those attending my seminars achieve perfect plots, proving that it pays you and me to double-check any plotting we do on log-log paper.

PROBLEM 2

The plot of the data is shown in Figure 6-12. Reading off the y value at unit 4 yields a cost of 120 hours (see circled points). Thus

$$LC\% = \frac{y_4}{y_2} \times 100 = \frac{120}{150} \times 100 = 80\%$$

PROBLEM 3

(a) An 85 percent learning curve is being generated.

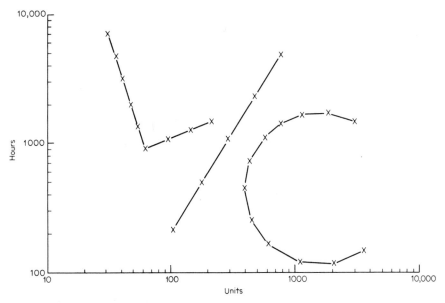

Figure 6-11 Answer to Problem 1.

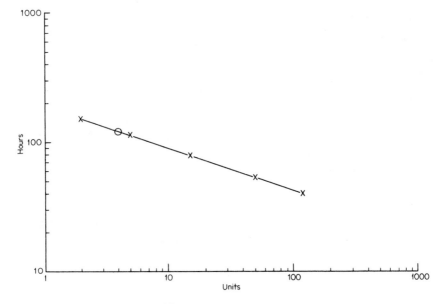

Figure 6-12 Answer to Problem 2.

(b) Slope constant $= \dfrac{\log 0.85}{\log 2} = \dfrac{-0.07058}{0.30103} = -0.23447$

(c) Calculating back from unit 15 gives for unit 1

$$a = \frac{250}{\text{antilog } [(\log 15)(-0.23447)]} = 471.74$$

(d) $y = 471.74(x)^{-0.23447}$

(e)

Unit no.	170	620	1400	3000
Hours	141.5	104.5	86.3	72.2

(f) $\dfrac{\log 0.90}{\log 2} = \dfrac{-0.04576}{0.30103} = -0.152$

(g) $a = \dfrac{180}{\text{antilog } [(\log 61)(-0.152)]} = 336.24$

(h) $y = 336.24(x)^{-0.152}$

(i)

Unit no.	170	620	1400	3000
Hours	154.0	126.5	111.8	99.6

PROBLEM 4

The data for least squares and coefficient of correlation calculations are

$$\Sigma \log x = 4.69020 \qquad \Sigma(\log x)^2 = 5.37020$$
$$\Sigma \log y = 12.16107 \qquad \Sigma(\log y)^2 \times 29.65713$$
$$\Sigma(\log x \, \log y) = 11.14299$$

(a) Cost of unit 1 = 487.42 hours.
(b) Slope constant = −0.27258. Thus the learning-curve formula is
$y = 487.42(x)^{-0.27258}$.
(c) It is a 82.78 percent learning curve.
(d) Predicted cost of unit 800 is 78.8 hours.
(e) Coefficient of correlation is 0.96.

These data are excellently fitted by the learning-curve formula, and the coefficient of correlation is very high. Yet even with such a good fit most readers will see differences between their eye-fitted line and the least-squares line. The differences may well be large, as the line was extended out to unit 800. If this occurred with a good correlation, as we have here, think of the differences that would occur when the points are much more widely scattered.

PROBLEM 5

There are only three estimating factors available in this situation: a 90 percent unit-cost learning curve should be applicable; you should be able to reach standard cost at unit 100; and standard cost for this black box should be 100 hours.

Since 50 units are one-half of 100 units, the cost of the fiftieth unit can be calculated as 100 hours (at unit 100) divided by 0.90, or 111.11 hours.

Since 25 units are one-half of 50 units, the cost of the twenty-fifth unit can, in turn, be calculated as 111.11 hours (at unit 50) divided by 0.90 or 123.46.

To calculate the projected cost of unit 500, a plot can be made of the calculated costs of unit 25, 50, and 100 and extended out to unit 500. Better yet, why not calculate it:

$$\text{Slope constant } b = \frac{\log LC}{\log 2}$$
$$= \frac{\log 0.90}{\log 2}$$

$$= \frac{-0.04575}{0.30103}$$

$$= -0.152$$

$$a = \frac{y_x}{\text{antilog}\,[(\log x)(\text{slope constant})]}$$

For unit 100 as x

$$a = \frac{100 \text{ hours}}{\text{antilog}\,[(\log 100 \text{ pieces})(-0.152)]}$$

$$= \frac{100}{\text{antilog}\,[2(-0.152)]}$$

$$= \frac{100}{\text{antilog}\,(-0.30400)}$$

$$= \frac{100}{0.49659}$$

$$= 201.37$$

Thus the learning-curve formula in this case is

$$y = 201.37(x)^{-0.152}$$

And for unit 500

$$y = 201.37(500)^{-0.152}$$
$$= 201.37(0.38883)$$
$$= 78.30 \text{ hours}$$

Because we want total hours at the various possible order quantities, it is easier to convert and use cumulative hours to arrive at the estimates needed:

Unit no.	Est. unit cost	Conversion factor		Cum. avg. cost		No. of units	Est. total hours needed
25	123.46	0.848	=	145.59	×	25	3,640 **(a)**
50	111.11	0.848	=	131.03	×	50	6,551 **(b)**
500	78.30	0.848	=	92.33	×	500	46,167 **(c)**

Notice that in this use of the learning curve you had to decide what LC should be achievable and at what unit number standard cost should be attained.

PROBLEM 6

Having calculated lot midpoints, we have:

Unit cost data		Cumulative avg. cost data	
Unit no.	Unit cost	Cum. pieces	Cum. avg.
5	1000	15	1000
25	686	35	821
50	583	65	711
90	508	115	623
140	458	165	573
215	414	265	513
315	379	365	476

The plot for this cost performance is given in Figure 6-13. On that same plot both the unit-cost and the cumulative average are extended out to 1000 units (see dotted lines).

(a) Readers can use their own plot, least-squares calculations, or a preprogrammed calculator. From the plot

$$\text{Unit } 5 = 1000 \text{ hours}$$

$$\text{Unit 10 cost} = 850 \text{ hours}$$

Thus

$$\text{unit cost learning curve LC} = 85\%$$

(b) Unit cost at unit 1000 will be 289 hours if the cost performance up to unit 365 is continued out to unit 1000.

(c) Cumulative average cost at unit 1000 will be 377 hours if the cost performance up to unit 365 is continued.

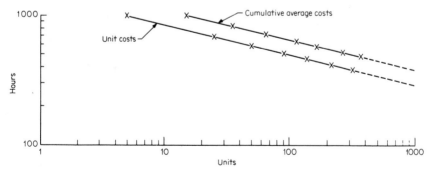

Figure 6-13 Plot for cost performance in Problem 6.

(d) Either line can be used for this solution. Using the cumulative average line, we know that to generate a cumulative average of 377 for 1000 units we must take a total of 377,000 hours to produce those 1000 units. We already know that it took 173,810 hours to produce the first 365 units. Therefore, the difference, 203,190 hours, is what it should take to produce the eighth lot of 635 units, or an average of 320 hours per average unit.

Using the unit-cost line, we see that the lot midpoint of the eighth lot of 635 pieces will be 365 + 635/2 or the unit cost at unit 683, which is 320 hours.

(e) Because we want to determine total additional costs, it is easiest to work toward this solution with the cumulative average line. We already know that the cumulative average cost for the first 365 units is 478 hours and that if this 85 percent learning-curve slope is continued out to 1000 units, the cumulative average of 377 per unit at 1000 units

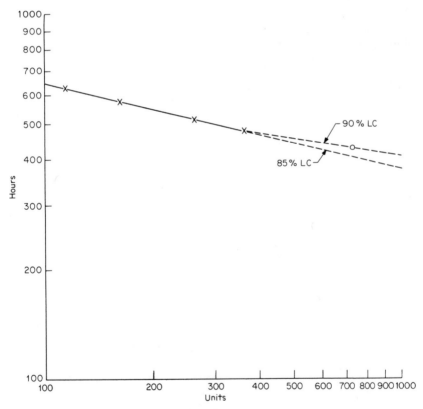

Figure 6-14 Answer to part **(e)** of Problem 6.

will be generated. Thus, a total of 377(1000) = 377,000 hours will be used to make the 1000 units. Therefore, we need now to calculate the effect of not an 85 percent learning curve but only a 90 percent learning curve for the eighth lot of 635 units.

Twice 365 units is 730 units. At a 90 percent learning curve the cumulative average cost for 730 units will be 90 percent of what the cumulative average was for 365 units, and (476 hours) (90 percent) is a cumulative average of 428 hours at 730 units. Having two points, we can draw a 90 percent learning curve after unit 365 and extend it out to 1000 units to determine the new cumulative average for that many units. This is illustrated in Figure 6-14, which is a blowup of the 100- to 1000-unit cycle on 1- by 1-cycle log-log paper.

Thus if only a 90 percent learning curve is achieved after 365 units through 1000 units, a cumulative average cost of 408 hours will be generated at 1000 units. Therefore:

LC%	Cost for 1000 units, hours
90	408,000
85	− 377,000
Additional cost	31,000
At $10 per hour	$310,000

Chapter

Specific Practical Estimating Techniques and Problems

This book is concerned with the management aspects of product-cost estimating, not with methods of costing any specific product. Actual estimating techniques suitable for a given product depend on the nature of that product and whether a conceptual or a detailed estimate has to be developed. For example, in detailed estimating on metal parts, the SME book edited by I. R. Vernon and the Hadden-Genger book listed in the Bibliography are excellent sources for approaches, formats to use, and even actual time values. In estimating in printing, box making, glass bottles, and similar products, well-defined estimating techniques are available as common knowledge and practice in the given industry. On the other hand, in other industries and products like electronics and heavy machinery and equipment, each company has tended to develop its own approaches, forms, and techniques. There is no one universal body of practice.

Certain specific practical estimating techniques that have applicability for a wide range of products and industries warrant mention though they are not applicable in *every* estimating situation. They may be overlooked in many situations where they could make a contribution. Certain estimating problems are brought up so often by those attending my estimating seminars that I discuss here likely approaches to their solution.

160

Quick-Set Labor-Cost Sheets

This valuable tool for building up a detailed estimate is frequently used by work-measurement people but less often by estimators. A *quick-set labor-cost sheet* is a one-sheet compilation of all the various elemental time values involved in the performance of a specific manufacturing operation. The work elements and their time values included on the sheet cover the vast majority of different types of units requiring that operation and also encompass the different sizes, weights, combinations of varieties of parts in subassemblies, and so forth. Developing an estimated labor time standard can be a time-consuming job: a lot of different standards have to be looked up in time-data books, tables searched out, and individual calculations made, to arrive at the total time needed for the given operation. If such labor-time data are condensed on one sheet, estimating goes much faster.

An Example In manufacturing electronics products a common operation is printed-circuit-board assembly, in which components like integrated circuits, resistors, capacitors, diodes, transformers, etc., are assembled on a printed-circuit board. On high-volume manufacturing automatic insertion machines are used, but even then some additional hand assembly may be needed. The printed-circuit-board assembly time must be estimated on a sheet like that in Figure 7-1. This particular sheet was developed for use at an electronics company with a wide variety of product lines, so that a variety of different printed-circuit boards had to be assembled, ranging from 20 to over 400 components.

The sheet in Figure 7-1 was a preprinted 8½- by 11-inch sheet. On the front, the side shown on the figure, are all the standard or most frequently used components on a printed-circuit board. As you can see, there are 14 different types. On the reverse side (not shown) are the time values of special work elements that are less frequently used.

The estimator working from the engineer's sketch of the circuit on the board or the schematic drawing of the circuit can read off the number of each of the different types of components, place them on the quick-set sheet, multiply quantity by the time values expressed in decimal minutes, total the base minutes, and calculate the standard hours needed per assembled board.

Obviously, such a sheet is a natural to computerize. The data and the calculating steps can be programmed on a minicomputer or even a microcomputer. Then the estimator feeds into the computer the numbers of the various types of components, and the computer applies the proper elemental time values and calculates total time required per board. One company doing this reduced the estimating time per board from 2 or 4 hours, depending on the complexity, to 10 minutes. In that

Quick-set labor standard sheet

Presolder PC board assy. operations — Printed circuit board dept.

Part No: _____ No. boards/cycle: _____ By. _____ Date: _____
Mfr. No: _____ No. of components: _____ Comp./sq. in.: _____
Operation description: _____
Tooling: _____

Element	No.	Time	Element	No.	Time	Element	No.	Time
Vertical lead comp:			**Integrated circuits:**			**Filter bandpass:**		
Bd. X comp. 1 2 3-4 5+			8 Lead	.53		5-6 lead – insert	.23	
— X 1 .15 .13 .11 .10			– socket	.10		**Crystals:**		
— X 2 .13 .11 .10			– socket assy.	.24		2 Lead insert	.10	
— X 3 .11 .10			9 Lead – 1 bolt, nut,	1.23		– solder	.20	
— X 4 .10			–2 bolt	1.76		**Resistor network:**		
Form 1 lead	.22		10 Lead	.54		5-7 leads-2 beads	.29	
" 2 "	.27		12 Lead	.54		8-9 " "	.40	
Tubing 1 – F & I	.39		– socket	.10		10-12 " "	.48	
" 2 – "	.49		– socket assy.	.24		13-14 " "	.55	
Axial lead comp:			8,10,12 – bend 4 leads	.75		**Sockets:**		
Bd. X comp. 1 2 3-4 5+			14 Lead – socket & bend	.32		Terminal-w/tweez'rs	.14	
X 1 .18 .17 .15 .14			" circ. to board	.29		Crystal – "	.07	
— X 2 .17 .15 .14			**Transistors:**			**Bare wire:**		
— X 3 .15 .14			3 Lead	.14		Cut, bend, insert	.24	
— X 4 .14			" – w/insul.	.22		" " , over comp.	.29	
			" – w/tubing	.99		**Insulated wire:**		
Coils and chokes:			" – w/3 spacers	.42		1 End – insert – bend	.10	
1 – 4 leads	.15		4 Lead	.15		" – solder – cup	.28	
Over coil form	.23		" – w/tubing	1.20		2 End (prop) – insert	.15	
	.12		**Transformer:**			" solder cup	.45	
Variable res. and cap:			Coil – 2 leads – fine	.31		Dress	.03	
2 Lead – long	.28		" 3 " "	.42		**Mount – demount:**		
" – w/tubing	.47		" 4 " "	.52		Open fixture	.09	
3 Lead – long	.21		" 5 " "	.63		Exp. holder	.23	
" – tubing 2	.54		Audio – 4 wire lead	.56		4 post holder	.11	
" – " 3	.63		2 Tab – 4 " "	.63		On table – from box	.07	
Pot – 2,3,4 lead	.13		Bent 6 " "	.66		Into slotted box	.18	
Stand off – 1,2 leads	.32		" 7 " "	.69		To table, box, nxt sta	.09	
			" 8 " "	.80		(From other side of page) Tot. special elements		
			I.F. Can – 4-6 lead	.13		Total base minutes		
			" " bend tab.	.21		Tot. base min. X 0.0192 = std. allowed hours		

Figure 7-1 Quick-set labor standard sheet.

particular installation the learning curve is also built into the calculation of the times required.

Many estimators do not yet have a computer available. A good starting step in the meantime is the quick-set sheet. If the company has a work-measurement group, as part of the production organization, it can be the source of such sheets. It is surprising that many such groups have not yet organized their own work-measurement data in such a format. They could save themselves a lot of time. In other companies the work-measurement group has such sheets for their use but the sheets are not yet used by Estimating. Estimators without a work-measurement group to provide the labor time data might still be able to organize the standards they have developed on their own in a format like the quick-set sheet. The labor involved will easily pay for itself over the many estimates they will be working on in the months ahead.

Other products suitable for quick-set sheets include the following:

All types of machining operations on metal parts (shearing, stamping, secondary punch-press operations, turning, milling, drilling, hobbing, broaching)

Foundry molding

Plastics molding and fabrication

Assembly operations such as small transformer assembly

A tremendous variety of subassembly work in a broad variety of industries

The technique is an old and valued approach that many estimators could use.

Estimating on the Curve

Another neglected tool is the *curve*. If you are estimating product costs in a situation where the same product is made but each order differs from the next in size, weight, configuration, etc., it is often possible to measure a correlation between product cost and one specific variable. If that correlation is good enough, the relationship can be mathematically stated and a formula developed to use for both interpolation and extrapolation of product costs.

An Example I had to estimate glass-bottle costs in Europe for a given time into the future. For a specific type of bottle (round, narrow neck) and a specific type of bottle-forming machine the estimates of total manufacturing cost (variable and fixed) were:

Bottle weight, grams	Estimated cost/ 1000 bottles
200	$ 47.30
260	56.50
400	77.80
575	107.85
740	140.05

A graph of the estimated costs by bottle weight is given in Figure 7-2.
Obviously, there is an excellent correlation in these estimates between the independent variable of bottle weight and the dependent variable of estimated cost per 1000 bottles. The straight line and three

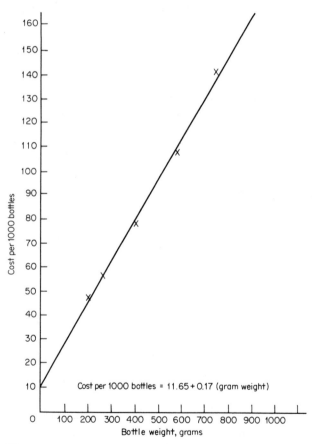

Cost per 1000 bottles = 11.65 + 0.17 (gram weight)

Figure 7-2 Projected bottle costs per 1000 bottles.

different curves were mathematically fitted to the five data points. The straight-line fit yielded the best coefficient of correlation, 0.9987, an excellent correlation. The straight line was therefore used to describe the relationship between the two variables and is also shown in Figure 7-2. With that line and its formula, the estimated costs of intermediate-sized bottles could be estimated. The line was also extrapolated to encompass bottle weight as low as 170 grams and as high as 920 grams.

In certain estimating situations such as this, it may be possible to identify a relationship between total product cost, including both variable and fixed costs, and a given product variable. In the above example, total costs were used, but in other instances and/or for other products, only variable costs per unit might be used. Such an approach has been used in plastics molding and die-cast products.

Another application of estimating on the curve is to specific costs. This kind of application has potential for many more estimators and many more product situations than the one described above. As before, only two variables are involved.

An Example Bottles are formed on interrupted-stage (IS) machines, of which there are numerous types. On an IS-6 machine 6 bottles are formed per complete machine cycle if it is single-gobbed. If the IS-6 is run double-gobbed, 12 bottles are formed per complete machine cycle. There are IS-5, IS-6, IS-8, and even IS-10 machines. Some can be run single-, double-, or even triple-gobbed. Thus there is a great variety of operation, bottle configuration, forming speed, and consequently cost. In this product and industry, forming speed is a great determinant of product cost per gross. Batch or material cost is also very important, but many costs are tied to forming speeds and forming-machine hours required per gross.

For a given size of IS machine double-gobbed and for a given bottle configuration, running speeds for various bottle weights are as follows:

Bottle weight, oz	4	5	8	10	11
Avg. bottles formed/min	122.1	109.7	96.0	81.2	85.2

The relationship between these two variables is shown in Figure 7-3. The straight line and three different curves were fitted to these five data points. The best coefficient of correlation, 0.982 in this case, was with the logarithmic curve formula $y = a + b(\ln x)$, where ln stands for the natural logarithm (base e). That curve is also shown in Figure 7-3. With the formula a table of forming-machine hours per hundred gross can be established for a range of bottle weights and used in building up estimates of glass-bottle costs.

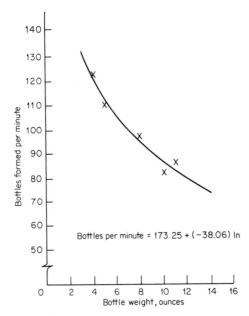

Bottles per minute = 173.25 + (−38.06) ln

Figure 7-3 Bottles formed per minute.

Such cost curves are widely applicable, but often they are not recognized, developed, or used where they could be. I have seen ingenious and helpful applications in such estimating situations as:

Plastics molding
Foundry pouring
Electronic subassemblies
Specific machining operations
Incinerator fire bricking

In estimating with the curve we are dealing with what the statisticians call *simple* or *linear regression*. (The word "linear" is used even though the relationship may be curvilinear, just as a straight line may be called a "curve.") Two variables are involved.

Lack of time is probably the major reason why estimating on the curve is so little used. Certainly it is not a technique that can be used in all product situations. Specifically, its implementation for estimating *total product* costs is usually limited to simpler product situations, but for more complex products, for specific cost factors, such as labor or machine cycle times, it has wide and realistic applicability in many product-cost estimating situations where it has yet to be used. To develop such curves for specific costs, an estimator must have time to assemble data, experiment with different product variables to detect usable correlations, and test various curves to find the best fit. But if the estimator can take this time, much time on many future estimates can be saved later.

Multiple Regression in Conceptual Estimating

Up till now talk of using multiple regression in product-cost estimating has been unrealistic. Multiple-regression analysis requires too much calculation and is too time-consuming to make it a suitable technique for most estimators, but increases in computer availability and capacity make the subject worth considering. It can be particularly valuable in conceptual or parametric estimating. A computer is not always necessary. With advanced hand-held calculators like the HP67, 97, 41c, or the TI59, multiple regression with two independent variables is possible. For more than two independent variables a computer is normally needed.

Multiple regression is a statistical tool that determines (1) the degree of correlation between two or more independent variables and a dependent variable and (2) the mathematical relationship between those individual independent variables and the dependent variable.

In product-cost estimating, the dependent variable is product cost, which usually depends on such independent product variables as size, weight, complexity by type of component parts, and so forth.

Consider costing multipart products, which can range from medium to high complexity. (The scope of such products is tremendous—instruments, controls, pumps, light machinery, heavy machinery, electrical products, electronic products in all their range, engines, and airframes.) It is common in such product situations for estimators to engage in conceptual or parametric estimating. Senior management and Marketing may want to consider the economics of a different but allied product or a new model of greater capacity or productivity. A customer may want an idea of the price your company would charge for an existing product, in a new model with better operating characteristics and higher performance. Since there are, of course, no detailed bills of material, much less detailed parts prints, a conceptual cost estimate has to be developed. On the other hand, some or many of the subassemblies and components in the new item will be like those being used in the present line. With good historical costs for both labor and material by type of subassemblies and/or components, an estimator can establish a quantitative relationship between total costs and individual categories of subassemblies and/or components. Then these measured relationships can be used to develop the conceptual estimate on the new proposed model.

An Example The company makes custom-built heat-treating equipment. Each order is for a unit that meets the customer's specifications for capacity, heat range, type of fuel, length of dwell, and other operating characteristics. Many orders are for one unit, fewer are for two units, and rarely are they for three or more. In responding to a request for quotation Estimating has no detailed bills of material or parts prints, but from the prospective customer's specifications Estimating can rough out in broad terms from experience such cost parameters as the following:

1. Overall unit size and weight
    a. Size and amount of heavy-metal angle stock
    b. Size and gauge of sheet metal
2. Type and amount of insulating material
3. Type and number of fuel-metering components
4. Type and number of electrical controls and components
5. Electrical-wiring harness size and complexity
6. Type and number of pipe fittings, connections, and lengths

Thus, there are seven major cost parameters or cost areas, which act as seven independent variables affecting the dependent variable of estimated total product cost per unit. The cost and importance of each of these seven parameters vary widely with the job. Estimating has actual material and labor costs for jobs completed in the past collected under these seven cost headings. Using those data, multiple-regression analysis can measure how variations in these seven cost areas in past actual jobs affected their actual final total cost. If a usable correlation is found, the mathematical relationships can be applied in the conceptual estimate for the new unit for which a quotation is being requested.

In the initial development stages in this example, the prospective job should also be estimated by the present conceptual estimating methods. Then the job's eventual actual total cost should be compared against the two estimates to determine the better predictor.

Notice the tremendous dependence on the quality of the past actual cost data and their identification or segmentation by cost parameter. This is true of all conceptual-estimating approaches I have ever known and used. If your production counts are poor, if your labor-time reporting is fictitious or not reasonably correct by category of work, you simply do not have good enough data for conceptual estimating on new jobs. Again we see the interdependence of Estimating and other management areas. This also applies to the important matter of estimating follow-up. If we have good estimating follow-up data so that we can make realistic comparisons of actual with estimated costs, we have a better basis for conceptual estimating when that is required.

Since such conceptual-estimating situations are frequently on long-lead-time products which may take months to manufacture, the necessary historical data extend over a number of past years. Because costs change over such a time, material costs will have to be indexed up to current costs. If labor times are available, they can be used as recorded unless major changes in labor efficiency occurred. But if actual labor records for past jobs are available only in money terms, they will have to be indexed to present actual cost levels.

Estimating under Conditions of Great Uncertainty

All estimating, involving as it does the task of predicting, entails some uncertainty, but occasionally estimators need to estimate a product cost under conditions of extreme uncertainty. Say that the company is contemplating a new product line that will advance the state of the art. Such a product may have some subassemblies or components like those in existing products, but major elements are completely different from anything made before. Obviously there can be no valid cost data for use as starting point. In another example, cost standards have to be developed for work where no standards have ever existed and for which no data are available. There are many other examples of estimating situations where there is no precedent or source for estimating data. All require estimating under conditions of great uncertainty.

In such a situation one technique to try involves the following steps:

1. Assemble around a table in a conference room, four, five, or six people who know the product and the cost areas involved.
2. Present them with the subject to be dealt with, a description of the product or part, or (if work standards are involved) the job to be done.
3. Have them all *independently* write down their estimate of the cost or hours.
4. Read out the independent estimates and determine the differences between them.
5. Together reconcile the differences and develop an agreed-upon Estimate.

An Example My consulting assignment was to improve delivery performance for a given type of product at a company producing electronic components. I found that one bottleneck was Production Engineering. All customer orders had to be processed through this department to translate the order into the company's part numbers and to have the needed parts specified and the operational steps determined. The easiest orders were repeats of past orders and simply had to be identified as such so that the data could be recovered from the files. Some required further study to decide on such parts as the proper connectors to use, the size of can to use, etc. A third type, the most time-consuming, required not only parts naming and sizing but also layout of detailed operational steps to make the product needed to meet the customer's specifications. A specific group of production engineers in the department handled this product as a full-time job. Thus orders fell into three categories on the basis of the work that had to be done by Production Engineering.

Although orders for this product had been increasing regularly, Production Engineering had added no new personnel. Senior management wanted quantitative evidence that additional people were needed and how many. Time studies were not a practical approach for economic and psychological reasons. The solution came from a technique used in job evaluation, where individual job factors for a given job have to be ranked with the same job factors for all the other jobs in the plant. This is done by getting managers together who know the jobs involved and having them rank the jobs independently. Their rankings are then revealed and differences reconciled.

In this case the department manager, the assistant department manager, the section head for the product, and his assistant, and I sat at a table. We agreed on the three types of orders that had to be handled. Then independently, we each wrote down for each type of order our individual estimate of how many orders a production engineer could process in a working day. We then read our estimates, defended them, discussed and argued about the differences, and agreed as a group on the final figures. In effect, we established standards for the Production Engineering effort on this product, and we could then give management the worker-hour workload and backlog.

In the example above the need was to develop work standards. The technique has been used just as productively for parts and product costs. When I discuss this technique at estimating seminars, I have been pleasantly surprised to find some estimators who have used it. Many of them observe how surprisingly close the first estimates, independently arrived at, are to each other. Only a minority of estimators have used this technique, however, and it has potential for many others.

This approach as used in estimating has been developed more formally by Andrew Van de Ver of the Wharton School. Called *nominal group technique* (NGT), it is described in two books listed in the Bibliography under Estimating in Uncertainty. The seven-step approach has been used successfully to develop programs and lay down project priorities. Although it has not been used for product-cost estimating, it has interesting possibilities in that area. Whether it is done on a more formal basis like NGT or the simpler steps given above, it could profitably be used in estimating situations where there is great uncertainty.

Another approach worth the reader's consideration is outlined by J. D. Burch (see the Bibliography). In the years to come, we can hope to see more written and published on real-life situations and the techniques successfully used to solve specific problems.

Paretoizing the Costs in Your Estimate

Many estimators often find themselves working under great pressures in time. Marketing or senior management need the estimate as soon as possible. As a result, estimators cannot take all the time they really need. Under such conditions, estimators should remember and apply *Pareto's law*, which has great applicability in all areas of management, not just in estimating.

Vilfredo Pareto (1848–1923) was an Italian economist and sociologist. Pareto's law states: *In any series of events to be controlled, a selected small fraction, in terms of numbers, will always account for a large fraction, in terms of effect.* For example, 20 percent of your customers will account for 80 percent of your sales, give or take some percentage points. It is a well-known inventory-management tool (ABC analysis of inventory) and applies to every company, no matter what the product or industry.

This law certainly applies to estimating because 20 percent of the parts or costs going into the product you are estimating will account for 80 percent of the total product cost. In most situations, estimators will know by experience which parts constitute the vital few. When you are working under great pressures in time, concentrate your limited time on the vital few. You can afford inaccuracies in the trivial many.

What Is Your Capture Rate?

The *capture rate* is the percent of total estimates made that result in customer orders. People often ask what a good capture rate is. There is no *one* answer. It depends on the product and industry.

In product situations and industries where the product is relatively simple and low in cost on a per unit basis and many estimates are made every week, the capture rate is invariably relatively low, e.g., printing, paper boxes, job-shop machining, stampings, and the like. In such product and industry situations, a 10 percent capture rate might be excellent. Many such shops average 5 percent.

With complex, multipart products, many of which may be custom products, the cost of developing an estimate is greater and the capture rate must be (and usually is) substantially higher. The time and cost involved mean that situations in which an estimate will be developed are more carefully selected. The number of potential customers is usually more limited, also. In such situations, a capture rate of 40 percent or more is common.

Two things should be said about the capture rate:

1. Keep track of your capture rate.
2. Have a good idea of what your competition is averaging.

Good estimates cost money to develop, and I have never seen or heard of an estimating group that had too much staff or time. In view of these facts, you cannot afford too low a capture rate, and you should keep track of what that actual rate is averaging. Surprisingly enough, many companies do not keep this record, which can be an excellent measure of how effective Estimating is in practice and how effectively the cost of estimating is being utilized. It is not a difficult record to keep and well worth the effort.

Because there is no *one* satisfactory capture rate, the best measure of whether yours is good or not is knowing the capture rate of your competition or other companies in your industry. This can take some digging, but it should be possible. Going to industry meetings, professional society meetings, and seminars and asking questions can yield some rough standards. If there are other companies in your industry in other geographical areas with whom your company is not in competition because of freight rates, asking direct questions can often determine what they are averaging on their capture rate. If you find that a like company is averaging 10 percent and you are only averaging 7 percent, some interesting and productive questions should arise on how you can make better use of your on-site estimating talents and your always limited estimating time.

How to Handle the Pseudo Customer

A provoking estimating problem is the pseudo customer for whom your group is always making a lot of estimates but who never or rarely gives your company an order. As one estimator put it, "Sometimes I think they are using me as their cost department." This can be a serious problem because the time wasted on these estimates can reduce the time available for more productive estimates and for improving the quality of your estimating. Probably the last thing you want to do is arbitrarily stop doing the estimates. First, you should try to use the situation as a selling ploy. Collect your facts; then have the salesman or representative go to that customer and say, "In the last year we developed 26 estimates for you, and not one order. We know we're more competitive than this. Don't you think this effort is worth at least one order?" Offer to go with the salesman to make the plea. Then if that does not work, you have to question seriously whether this company will ever be a customer.

This can also be a delicate problem, on two counts. (1) Estimating will be bucking Marketing, who tend to figure that it never hurts to bid. But the bid requires an estimate, and that costs money. (2) In breaking into a new customer, a long time and a lot of bids may have to be gone through before the door opens, but eventually it does. One estimator told of a customer for whom a great many estimates were made, who came through with an order only after years but now was one of the company's best. The whole problem requires a good Estimating-Marketing interface and good communications. With these it is usually possible to determine whether the "customer" is worth the continued effort and cost.

How Accurate Should the Estimate Be?

Like the capture-rate problem, there is no *one* answer to this problem, which depends on the type of product and three interdependent product characteristics:

Complexity of the product
Individuality or differences between orders for the product
Uniqueness of the product

The starting point for any answer to this problem is knowing your present actual accuracy. This obviously requires good estimating follow-up data so that you can compare actual costs with the original estimates. Only when companies have a good follow-up data collection and comparison capability can they really know their present estimating accuracy.

Also when we consider accuracy, we should be concerned as much with favorable variances, or estimates higher than actual costs, as we are with unfavorable variances, or estimates lower than actual costs. The former can reflect sloppy estimating with too much insurance built into the estimate. These clearly can have had a poor competitive effect on the product price developed from that estimate, something managements are not always sensitive to. Most emphasis is usually placed on the degree to which actual costs exceed estimated costs.

Generally, the simpler the product, the lower the variance between estimate and actual should be. Thus in toiletry products, boxes, stampings, printing, and like items a ± 2 to ± 3 percent accuracy should be the goal. If the variance is more than ± 3 percent, it is usually profitable to spend money to correct the situation. In large, complex multipart products a ± 10 percent accuracy may be an operating reality and quite competitive.

A related factor is the degree to which there are differences between orders. Consider two companies in the same industry. One has a high proportion of orders and estimates for standard or stock items. Another has a much higher proportion of orders for special or custom items. On the average estimating accuracy should be higher in the first company. The second company's estimators have more variables on which to predict costs.

In conceptual or parametric estimating the range of expected accuracy has to be higher and wider. When you are dealing with state-of-the-art situations on large complex products, a ± 25 percent accuracy may be as good as can be expected. Again the product complexity has to be a factor. You might be advancing the state of the art in a less complex product situation where ± 15 to ± 20 percent should be achievable.

Many guesstimates have been offered on this subject. Some ranges of ± 40 percent have been advanced on conceptual estimating on large complex products. That seems too loose a standard to me. Although airframes are large, complex, and entail new types and higher specifications on new models, there is one famous airframes company who would never tolerate a ± 40 percent estimating variance.

The first criterion is doing better than your competition. Unlike the capture rate, it is much more difficult to determine your competitors' estimating accuracy. As a result, you have to set your own goals and then meet them and—better yet—gradually improve them further.

Size of the Estimating Staff

How large should the Estimating staff be? This is one of the great unanswerable problems of the estimating field. In any given situation it depends on:

The product and its complexity
The number of estimates required per month

The proportion of custom jobs, or "specials"
The strength and rigor of the competition
The quality of the estimating talent on hand
The data, services, and facilities made available for Estimating's use

I have never seen an overstaffed Estimating department, but I have seen more than a few which were understaffed and which could have improved their estimating performance with more personnel.

The greater the complexity of the product, the larger the Estimating staff needed. Even with relatively few estimates per month, when the product has many complex parts, many subassembly levels, and many purchased components, the estimating hours per bid are great and a larger staff is needed.

For simpler products, the determinants of staff size are the number of requests for quotations *and* the proportion of requests that are for custom models, specials, or one-offs. In the glass-bottle industry a company with a steady list of customers and on-going orders can have computerized estimating handled reasonably well as a part-time job in Cost Accounting. In medium-sized job machine shops, printing houses, and box shops, the variety of different orders and the sheer number of requests for quotations demands a full-time estimating group. In a given industry or area, some companies have historically marketed a greater proportion of custom jobs, specials, or one-offs. Such a company will need a bigger estimating staff than one concentrating on a standard, more limited line.

Normally, the greater the competitive pressures on price, the greater the need for better estimates and the more estimating hours needed per year. Because of market position, capacity or facilities capability, or patent protection, some companies can afford less accurate and less careful estimates and thus can get away with a smaller Estimating staff. Such situations can be dangerous because they can lead to complacency and management inertia. Suddenly new, brighter, and sharper competition may appear, and too late comes recognition that a larger and better estimating effort is needed. If response time from competitors is relatively short, faster estimating and a larger Estimating staff is normally needed.

As in every management area, the quality of the talent being used in Estimating has a great effect on the staff size needed for a given workload. If the Estimating management and the talent in the group are above average, they will develop techniques and tools that facilitate their work. Equally important, they will always be making moves to shorten the time involved in their estimates. As a result, they handle more estimates and turn out higher-quality estimates than a less talented group can. This is why it is so sad to see a company lose good estimators because they do not pay them enough.

A factor of great weight in determining the size of the Estimating effort is the type and quantity of data, supporting services, and facilities available for Estimating's use. If, for example, the company has a work-measurement group, labor standards are available and Estimating does not have to develop such standards on their own. If there is a Manufacturing or Production Engineering department, they can supply parts routings and perhaps tooling-cost estimates as well. If Estimating has access to computer facilities, more work can be handled by the same estimating staff.

Obviously for one given Estimating group and situation a great many variables affect the size of the staff needed. Company management has to arrive at its own answer for its own situation. The answer will depend on affirmative answers to questions like these:

Is Marketing reasonably satisfied with the response time of the Estimating group? Is it better than our competition's?

Is our capture rate better than our competition's?

Are our Estimates of satisfactory accuracy as determined by the follow-up comparison of actual with estimated cost?

Are our estimates consistent?

Are we estimating today as we did 3 years ago, or is steady progress being made in our estimating techniques and the facilities available for Estimating's use?

Estimating and Price Analysis

A very valuable interface between Estimating and Purchasing warrants mention because it is often underutilized and even ignored. Purchasing can use the talents and knowledge of Estimating to estimate the cost to manufacture products they are buying from vendors. Knowing the cost to produce, Purchasing is better able to negotiate the final price intelligently. The technique has been referred to as *price analysis*.

For example, in a metalworking company, if Purchasing is buying metal parts, Purchasing's associates in Estimating should be able to study the parts being purchased and come up with a good estimate of how much it should cost the vendor to make them. Estimating's time limitations in real life, of course, allows this to be done on only major items involving a lot of money. But suppose that tens of thousands of dollars are being spent in the course of a year on a given metal part. The development by Estimating of an estimated production cost for that part can be a profitable use of Estimating's time and talent if, as a result, Purchasing can drive a harder bargain and still leave the vendor with an adequate profit. Vendor costing and subsequent pricing often leave much to be desired; the items your company is buying may be helping to carry other items the vendor is selling to other customers. You lessen the chances of this happening if Purchasing has a good estimate of the production cost of the items they are buying.

Purchasing often has difficulty in getting vendors to reveal their production costs in any meaningful detail or even to discuss production costs at all. When a vendor is confronted with a detailed estimate of the cost to make the item it is trying to sell you, it may dispute the estimate, but to do so, it has to get into cost details—material, labor, and the overheads it is applying. Then a lot can be learned and perhaps some illogical costing on the vendor's part identified.

Some companies use this price-analysis approach most effectively on every major item they can. In fact, it is standard operating procedure. In other companies where it is equally possible the idea is never broached.

Chapter

Estimating in Make-or-Buy Decisions

Most cost estimators, at some time or other, find themselves involved to some extent in make-or-buy calculations and decision making. There is a tremendous range between companies in their practice on make-or-buy in such matters as:

Sought-for opportunities versus emergency make-or-buy studies

The management functions which do the studies

The number of data applied in the studies and calculations

The logic followed in the studies to reach the final decision on whether to make or to buy

In dealing with make-or-buy, the approaches, practices, and logic followed in real life are so varied that this chapter starts off with a case problem. Readers are urged to do this problem before proceeding further into the chapter or looking at the answer. The problem will clarify how your training, approach, and logic affect your decision making in this important area of make-or-buy.

Case Problem in Make-or-Buy

Assume that you are the owner of the company. This mental orientation means that you can follow your own approach and logic and make

178

up your own mind about the proper decision. It's your money and your responsibility. (Of course, as managers, we should have this orientation whether we own the company or not.)

The Quality Instrument Company, which you own, manufactures a proprietary line of instruments. The company does all its own assembly and testing and much of its own parts fabrication. For example, most of the metal parts used in the instruments are made in the plant's good-sized machine shop, where a wide variety of metalworking operations are performed. The department has automatic screw machines, shears, punch presses, lathes, mills, and drill presses, with an average 80 per-cent machine utilization and an average 5 percent scrap. Average hourly direct labor cost in the department is $6 per hour, and labor efficiency averages 100 percent of standard. The company operates only one shift, and that is all you want to operate. The department supervisor is a thoroughly experienced machinist and manager with many years of experience in a broad range of metalworking operations.

One metalworking operation the company has never done is gear hobbing. The company has always bought the three gears used in its major instrument line. These gears are used as sets with one gear each per set. The forecast yearly sale is 16,800 instruments, and while cosmetic changes are made from time to time, the basic parts, including these gears, are not up for redesign. You have every expectation that the instrument will be in your line in the years ahead. Each gear takes 2 ounces of metal costing $4 a pound.

A brief description of the operations required and of gear hobbing may be helpful. The gears average approximately ¼ inch in diameter. The starting operation is to make the gear blanks on an automatic screw machine from bar or rod stock of the metal used. The screw machine will turn the two diameters of the gear blank, drill the center hole running through the blank, and cut off the gear blank from the bar or rod. The result is a finished gear blank, ready for the next operation. The gear blank then has teeth cut on the blank's largest diameter on a machine tool called a gear hobber. It's about the size of a small desk. The machine operator places a gear blank in the machine between holding centers, and in the machine cycle a hobbing cutter automat-ically cuts the teeth. An operator can run two, three, or even four machines on this class and size of work. The cutting action and accu-racy are all built into the machine and no lengthy training period is required for this type of machining. Any given hobbing machine can cut any one of the three gears involved. To change the setup from one gear to another, the hobbing machine's internal gearing and the hob-bing cutter used are simply changed. Occasional burring is needed after hobbing; it can be done by the hobbing-machine operator while overseeing the operation of the machine or machines.

The controller and cost accountants have developed the manufacturing burden as 250 percent of direct labor. They estimate that fringe benefits are 25 percent of payroll in addition to the 250 percent manufacturing burden. Also, they like to assign a 10 percent overhead to the value of purchase orders to cover Purchasing department and purchase-order writing costs. They depreciate the average machine tool over an 8-year life, but for profit and loss and tax calculations an accelerated depreciation plan is used, under which, for all practical purposes, 25 percent of the new equipment cost is charged to the first year of operation, 25 percent of the remaining value to the second year, and so forth, until by the eighth year, the asset value is very small and is then completely written off. Setup labor is classified and accounted for as indirect labor and is included in the 250 percent manufacturing burden.

Sufficient automatic-screw-machine capacity, with each operator running three screw machines, is available in the machine shop to make the gear blanks out of bar stock. There is sufficient floor space open in the machine shop for some new equipment such as gear hobbers but of course, the equipment must be bought. Manufacturing Engineering has determined that the type of gear hobber needed costs $32,000 each with an eventual salvage value of $1000. Costs to install each machine are estimated at $1000. Hobbing cutting tools cost $300 each; with four regrindings or resharpenings possible in a tool's life, each tool, over its life, will hob, on the average, approximately 4,000 gears. Additional facts are shown in Table 8-1.

The other case facts can be summarized as follows:

Machine shop facts	100% labor performance to standard
	5% scrap as overall average
Yearly requirement	16,800 sets/year
Material	$4/pound
	2 ounces of metal needed/gear
Labor cost	$6/average hour
	25% fringe benefit costs
Tooling cost	$300 purchase cost for a hobbing cutter
	Four regrindings possible over a tool's life
	4000 gears produced over a hobbing cutter's life
Accounting overhead calculations	10% on purchase orders
	250% on direct labor costs

Thus you have data on just four elements of cost, material, labor, the overhead you think applicable, and tooling. With your calculator, estimate the cost required to make a set consisting of one each of gears A,

TABLE 8-1

| | Time, minutes† | | | |
Gear	Forming blank	Hobbing and burring	Regrinding or sharpening hobbing tool	Present purchase price
A	0.50	3.00	60	$1.25
B	0.50	3.25	60	1.50
C	0.50	3.75	60	1.75

† All times are standard times and include necessary personal, fatigue, and delay allowances.

B, and C. Then decide whether you should make these gears or continue to buy them.

Readers are urged to do this first on their own, completing the estimated total make cost for each gear and thus per set. In the answer, which follows, I play the role of the owner. But doing it first yourself may well be enlightening.

Solution to Make-or-Buy Case Problem

Before examining this solution, the reader is referred to Chapter 4, where in the section Problems with Conventional Absorption Costing the case of the transformer manufacturing department in the electronics company was related. It will be referred to later in supporting the logic used in the following solution.

MATERIAL COSTS

2 oz metal/gear at 10% scrap $= \dfrac{2\ oz}{0.90} =$ 2.22 oz metal/good gear made

$\dfrac{4/lb}{16\ oz/lb} =$ $0.25/oz

(2.22 oz/gear)($0.25/oz) = $0.556 material costs/good gear made

Credit allowance for scrap is unknown but very small.

Notes

Average scrap is 5 percent, but scrap will occur not only in the gear hobbing but also in making the blanks on the automatic screw machine. Therefore on the first of the two machining operations required the scrap allowance is doubled. This is a conservative approach that increases the calculated make cost.

There is a 10 percent overhead on the value if the purchased material is not used. This might have a certain logic in inventory management but for make-or-buy decision making, it strikes me, as the owner, as not sensible or realistic. If I make these gears, instead of buying them, the company will not need another buyer, more receiving labor, or more of any of the costs normally included in such a purchase order burden. Labor costs are shown in Table 8-2.

Notes

Again, 10 percent scrap is built into the automatic-screw-machine labor costs, to allow for 5 percent scrap there and 5 percent scrap on the subsequent hobbing operation.

The 0.68 factor used to convert hobbing-machine minutes into labor cost is derived by dividing the total machine hours needed in the year to make 16,800 sets of gears by the 2000 worker hours the new hobbing-machine operator will work and be paid on one shift (computations below). The effect of this is to charge all the cost for the new operator's full year to these gears. It is a conservative approach that accepts less than perfect production schedule, the hobbing operator

TABLE 8-2
Labor Costs

$$(\$6/h)(125\% \text{ to include fringe-benefit costs}) = \frac{\$7.50/h}{60 \text{ min/h}} = \$0.125/\text{labor min}$$

	Gear A	Gear B	Gear C
Automatic screw machining (make gear blanks):			
0.50 min/blank			
0.90 (10% scrap)			
(0.556 machine min/blank)($0.125/min)(0.33)			
(for 3 machine operations)	$0.023	$0.023	$0.023
Hobbing (cut gear teeth):			
(A) $\dfrac{3 \text{ min/gear}}{0.95 \ (5\% \text{ scrap})}$			
$\quad = (3.158 \text{ machine min})(\$0.125/\text{min})(0.68 \text{ partial}$			
$\qquad \text{2-machine operation})$	0.268		
(B) $\dfrac{3.25 \text{ min/gear}}{0.95 \ (5\% \text{ scrap})}$			
$\quad = (3.421 \text{ machine min})(\$0.125/\text{min})(0.68)$		0.291	
(C) $\dfrac{3.75 \text{ min/gear}}{0.95 \ (5\% \text{ scrap})}$			
$\quad = (3.947 \text{ machine min})(\$0.125/\text{min})(0.68)$			0.336
Total direct labor cost	$0.291	$0.314	$0.359

sometimes having only one hobbing machine to operate. As seen below, 2940 hobbing-machine hours will be needed to make these 16,800 sets of gears. To be more optimistic you could calculate that only three-quarters of the new operator's time in the year would be needed, and the other quarter could be used on other work and not charged to these gears. This would reduce the hobbing labor costs included in the make costs. But, again, let us be conservative.

The 250 percent manufacturing burden is not used. That burden includes the plant manager, the building itself, supervision, maintenance labor and supply cost, Production Planning and Control, and all the other cost matters normally included in manufacturing overhead. If I make these gears, instead of buying them, I am not going to need another supervisor, added people in Maintenance or Production Planning, or a larger building, etc. So as the owner I see no sense in applying that burden percentage. Obviously, this point is critical to the entire case and the logic being applied is more strongly defended after the entire make cost has been calculated. So bear with me until then.

However, it is only fair and logical to include in the make costs of these gears any *additional* overhead that will be *newly* incurred if it is decided to make them (what Accounting calls *marginal overhead*). Such additional overhead would be the depreciation on hobbing machine(s) which the company does not own now. This leads us to the question of how many hobbing machines we would need to buy.

ADDITIONAL OVERHEAD INCURRED

$$3.00 + 3.25 + 3.75 \text{ min} = \frac{10.00 \text{ min/gear set}}{0.95 \text{ (5\% scrap)}},$$

$$= \frac{10.526 \text{ min/set}}{60 \text{ min/h}} = 0.175 \text{ machine hour/set}$$

$$(0.175 \text{ machine hour/set})(16,800 \text{ sets/yr}) = 2947 \text{ machine hours/yr}$$

$$\frac{2947 \text{ machine hours/yr}}{2000 \text{ working hours/shift-year}} = 2 \text{ machines needed}$$

Thus the labor factor for multihobbing machine operation is

$$\frac{2000 \text{ worker hours/yr}}{2947 \text{ machine hours/yr}} = 0.68 \text{ hobbing labor factor}$$

$$(2 \text{ machines})(\$32,000/\text{machine} + \$1000 \text{ installation} - \$1000 \text{ salvage})$$
$$= \$64,000$$

$$\frac{\$64,000}{8 \text{ yr life}} = \frac{\$8000/\text{yr}}{16,800 \text{ sets}} = \$0.476/\text{set} = \$0.159/\text{gear}$$

Notes

For tax and profit and loss application, I would want the controller to accelerate depreciation costs. But for make or buy, it is an accounting artificiality. If I follow that logic, I end up calculating that the gears will cost me so much more the first year than the second, so much more the second year than the third, and so on. In this solution, depreciation is therefore straight-lined over the acceptable 8-year period.

The reader will have noticed the deliberate oversimplification in the above calculation. The $1000 installation cost is offset by the $1000 salvage value of the machines. This is wrong, because it ignores the time value of money. The installation cost would be out-of-pocket when the machines are installed, but the salvage-value income is 8 years away, and $1000 eight years away does *not* equal $1000 paid out today. If you put that $1000 salvage income in your financial calculator, at say 10 percent (at monthly compounding) over 8 years, it is worth $370 today. But this decision will not swing either way for a cost of

$$\$630(2 \text{ machines}) = \$1260$$

HOBBING TOOLING COST

$300/hobbing tool + [4 h regrinding(sharpening)]($7.50/h)

$$= \frac{\$330/\text{tool}}{4000 \text{ gears/tool life}} = \$0.083/\text{gear}$$

Summary

	Gear A	Gear B	Gear C	Total set
Material	$0.556	$0.556	$0.556	$1.668
Labor	0.291	0.314	0.359	0.964
Additional overhead	0.159	0.159	0.159	0.477
Tooling	0.083	0.083	0.083	0.249
Total make cost	$1.09	$1.11	$1.16	$3.36
Buy cost	1.25	1.50	1.75	4.50
Saving	$0.16	$0.39	$0.59	$1.14
Saving/year for 16,800 sets				= $19,000

Thus, as calculated, the company would be $19,000 ahead if it were to make these gears instead of buying them. If the company were making 10 percent pretax profit, the result of making these gears would be equal to the profit on closing a $190,000 sale.

The entire decision hinges on whether or not we include that 250 percent manufacturing overhead. As I am calculating make costs, the situation is $3.36 make cost versus $4.50 buy cost, for a very clear decision to make. Now, let us follow the widely used approach and apply that 250 percent manufacturing burden. To do it most simply, we can multiply the $0.964 labor cost per set by 250 percent and thus add an additional $2.41 to the calculated make cost. Then make cost equals $3.36 + $2.41 = $5.77 per set for an obvious buy decision. Thus we are faced with two choices, 180 degrees apart:

$3.36 versus $4.50 → make

$5.77 versus $4.50 → buy

If I were the owner, I would make these gears. To defend the logic applied, allow me to advance the following scenario. Suppose I make these gears and the make costs work out pretty close to my calculations. Then 2 years later, my controller comes to me and says, "Boss, I've made a study of those gears we're making, and at our costs, a set of gears costs $5.77 to make. And we can buy them for $4.50." I listen to his logic and buy the gears. And I end up with two idle hobbing machines on my machine shop floor. Isn't that exactly what happened in the transformer department in the electronics company mentioned in Chapter 4? If that logic does not make sense to you 2 years from now, it cannot make sense today. And isn't it a pity that at that electronics company some manager in Estimating or in Purchasing when the decision was made to buy that first large-volume transformer item didn't ask, "If you buy that transformer, what do you think will happen to the overhead that work is absorbing? Do you think it will all just disappear?" Perhaps they did, but senior management did not listen. I would like to think that Estimating raised the question emphatically.

We have all been so long and carefully indoctrinated with the classic approaches to overhead applications that most managers and Estimators will calculate make costs that include the manufacturing overhead and decide to continue buying. A definite minority ignore the manufacturing overhead and thus decide to make. Most of the majority group eventually agree with the logic advanced here and agree to make, but before accepting the make decision, they must first hurdle the difference between make-or-buy decision making and product costing under conventional absorption product costing. They maintain that these gears, like the rest of the parts in the instrument, have to bear their fair share of the manufacturing burden. Of course they do, but that is product costing. In this case, if we made these gears, Cost Accounting would have 2000 more labor hours over which to spread the manufacturing fixed costs. As a result, the 250 percent would be reduced to, say, 248 percent. Then in costing the instrument under conventional absorption costing, all the parts, including the gears, would have 248 percent manufacturing burden applied against the labor included in their manufacture.

Many people attending estimating seminars point out that their company's controller would insist on their including manufacturing burden in the make calculation because it is company policy. If company policy does not stand up to the cold, harsh light of logic, if it results in underutilizing company capacity or even emptying a department of work, or if it denies us an opportunity to add $19,000 to yearly profit, should it be followed blindly? Higher management may not listen today, but Estimating has the job of advancing their ideas and logic and continuing to advance them. Ideas that make basic good sense will eventually prevail.

To sum up the practice proposed and exemplified in this case:

When calculating make costs, you should include in those costs only that *additional* overhead (often called *incremental overhead*) that will be *newly* incurred if you decide to make.

And the corollary: when you are comparing a present make cost with a proposed buy cost, you should include in that make cost only those overhead costs which will be identifiably *eliminated* or saved if you decide to buy.

The Range of Make-or-Buy

Make-or-buy decision making covers a broad spectrum the extremes of which might be identified as *unit-cost comparison* and as *business modeling*.

The case problem at the start of this chapter is an example of unit-cost comparison. We know what the gears cost us per unit to buy, and our task is to calculate the cost per unit to make. Knowing these costs, on a per unit basis, we can compare them and make our decision. This is the simplest category of make-or-buy decision, at the easier end of the spectrum. It entails parts or items related or similar to those already being manufactured by the company. Sometimes, as in the case problem, more or different production machinery may be needed, but basically, the work involved is simply an extension of the type of manufacturing already being done. Many companies farm out such parts in busy times and take them back in slower times. For such items, a unit-cost comparison is usually the practical approach.

At the other end of the make-or-buy spectrum is the much more complicated and difficult task called business-modeling make-or-buy. Here we may be dealing with existing or new items or products different from anything the company is making now. It may be the equivalent of getting into a new business. Business-modeling make or buy almost invariably includes more complex factors, including many intangible factors to which a money value cannot readily be assigned. In effect, a model of the hypothetical business may have to be made.

Assume that your company is a famous food manufacturer. You have plants all over the country and in the course of a year you buy millions of glass bottles from many suppliers. Your director of packaging has an idea that management wants to investigate. Suppose you build a glass-bottle plant of your own. A single plant would supply only a small percentage of your total nationwide yearly needs, but with such a plant (1) you would have a quality standard of your own manufacture and (2) you would know the cost to produce a gross of bottles. When you are buying hundreds of thousands of gross, that is a good thing to know. Two early decisions you would have to make are (1) how big a plant and (2) where, because both size of plant and location obviously affect operating costs. If you decide on 4 possible sizes as measured by numbers of gross to be produced and on 5 possible locations, you are really talking about 20 possible businesses.

For each of these 20 possible plants, a projected profit and loss statement should be developed. The income projection is easy because you know what you are now paying per gross of bottles at those locations. The next task is to project or estimate the costs to own and operate each of these 20 possible plants. You have to estimate costs for material, direct and indirect labor, equipment and depreciation costs, fuel, power and other utility costs, the costs of spares and supplies, and all the other costs. In addition, to be realistic, you have to include start-up costs, and on something like that cost you are dealing with a matter of vast uncertainty. But you need to estimate all the costs involved because your decision will depend first on whether any of these possible plants would be profitable. Only after you have determined projected profitability can you calculate return on the needed investment. Such business modeling is a matter of many factors, great complexity, and many intangibles, but it is still make or buy.

Most make-or-buy analyses in real life are unit-cost comparison. Many writers on the subject of make or buy criticize such comparisons as too superficial and not giving enough weight to nonquantifiable factors that should bear on the decision. Inevitably, however, they include such comparisons in their recommended procedures because such comparisons offer the clearest basis for decision making. After all, management needs some facts and quantifiable measures upon which to decide. It is essential, however, that the make-or-buy analyst always be sensitive to *all* the factors that should be included in such a unit-cost comparison, including factors which should be considered but which cannot be stated in money terms.

Opportunities for Make-or-Buy

Probably 75 to 80 percent of the components you are buying you should be buying, and probably 75 to 80 percent of the components you are making you should be making. But the fact that the complementary percentage, 20 to 25 percent, at most, is relatively small, does not mean that there are not attractive opportunities for make-or-buy analysis. In my experience, the potential can lie in both new and old products. Decisions made on new products often blindly follow precedent, and decisions on existing products are often precipitated by crisis. Let me defend both statements.

On new products, someone has to determine whether to buy or to make each of the components going into that product. Who makes that decision differs from company to company (design engineers or manufacturing or production engineers). Whoever it is comes to a part of a kind the company has always bought and instinctively marks it "buy" or a part of a kind the company has always made and marks it "make." They are blindly following precedent. For most parts, this will probably be correct, but some opportunities may be missed as well.

On existing products make-or-buy analyses are often made only because there is a crisis. For example, you are buying an item you could make. The vendor starts creating problems in quality, delivery, and/or price. The problems become so acute that finally management decides to have someone compute the cost to make. As another example, the shop load and backlog are climbing sharply, the plant is overloaded, and late deliveries are increasing. Faced with these problems, management orders studies to see which parts can economically be subcontracted. Again, a crisis precipitated the make-or-buy analysis. Things are happening to you; you are not making them happen.

This blind following of precedent and reaction to crisis occurs too often to conclude that all the make-or-buy opportunities are being discovered and exploited. Even if a serious and coordinated make-or-buy analysis effort has been made at the design stage, it is only realistic to expect some opportunities to have been missed. Processes and designs change; operating conditions change. As a result, previously sound decisions on new or old products may no longer apply. The make-or-buy analysis should be a continuing effort in most companies. It offers a real opportunity for Estimating to increase its effectiveness by detecting chances for profit improvement, but the identification of make-or-buy opportunities can come from any number of management areas. The most obvious are:

Process or Manufacturing Engineering
Purchasing
Estimating
Industrial Engineering
Production Planning and Control
Cost Accounting
Production management

Many companies with a continuing make-or-buy analysis program use a make-or-buy committee. Its effectiveness will vary directly with the strength and ability of its leadership, which is usually Manufacturing Engineering, Industrial Engineering, or, less frequently, Purchasing.

We should not forget the Pareto effect. In our planned search for make-or-buy opportunities, we should be looking for the vital few to avoid dissipating our limited time on the trivial many. Since we should put our time where the money is largely spent, the first task is to identify those parts which not only offer a make-or-buy opportunity but also entail the most money in a year's time.

Estimating's Role in Make-or-Buy

Product-cost Estimating's responsibility and the techniques it uses make it a natural for an active role in the make-or-buy analysis.

In building detailed cost estimates, Estimating must review each type of part and its present status as a make or as a buy. Knowing the operation as they must, they can identify profitable opportunities to make what is now being bought. Conversely, knowing the practical realities of operations and operating costs, they can detect potential buy possibilities that it may be profitable to investigate.

As observed above, the opportunity for a potentially profitable make-or-buy analysis can come from any one of a number of management areas, which may well calculate the make costs. Thus, if the item is presently being made, the make costs have to be determined, and many of the inputs needed will be obtained from Cost Accounting. If the item is presently being bought, the make cost has to be calculated, normally by the originator of the idea. In either case, product-cost Estimating has a contribution to make. Estimating should at least review and critique the make costs calculated by Cost Accounting or the make cost predicted by others.

The most common error made in make-or-buy analysis is misapplication of overheads. The second most common error is to be too optimistic on the make costs calculated. This error is particularly likely to occur in calculating make costs on an item presently bought. Estimating must be the on-site, hard-headed realists on product costing. As such, they have an important contribution to make in reviewing and criticizing the make costs calculated by other management areas.

Factors Included in Make-or-Buy

Besides being done blindly following precedent and/or only in a crisis situation, make-or-buy decisions are often made with too few data. This may be caused by time pressure, lack of usable historical costs, or failure to consider all the cost factors involved. At Estimating seminars, the case problem at the beginning of this chapter is given. After participants have developed their make costs and made their decision and we have reviewed and discussed my solution, I ask whether in their work they don't often find themselves making make-or-buy calculations and recommendations with *fewer* data than they were given in the case problem. The majority respond that they do. Certainly when we make decisions involving thousands or tens of thousands of dollars, we should make every effort to collect and use all the applicable data we can get. If they are not all we should have, we must take steps to have more of the necessary data in future make-or-buy situations. This may seem obvious, but some companies have no better data today on which to base make-or-buy analysis than they had 5 or even 10 years ago.

The factors involved in a given make-or-buy analysis will depend on the value and complexity of the particular item under consideration, but specific factors to be included can be classified into quantifiable and unquantifiable factors.

QUANTIFIABLE FACTORS

These are the factors that we should be able to state or estimate in dollar, time, or percentage terms. They are the more obvious factors and can include:

1. Raw-material cost
2. Scrap percentages as they affect material and labor costs
3. Learning costs at the start of production

4. Freight in and incoming inspection on both raw material and purchased parts
5. Labor needed
    a. Cost per hour
    b. Productivity per hour
    c. Cost per unit
    d. Overtime
    e. Incentive pay if applicable
6. Fringe-benefit costs on labor
7. Additional overhead because of additional equipment needed
    a. Depreciation based on:
        (1) Acquisition cost
        (2) Salvage value
        (3) Installation costs
        (4) Realistic estimate of equipment life
    b. Maintenance costs
    c. Spare parts
8. Tooling costs
9. Power and/or fuel costs
10. Additional supervision, if needed
11. Inventory-carrying costs if additional inventory is involved
12. Additional space, if needed
13. Cost of money tied up in space and equipment
14. Material handling costs

UNQUANTIFIABLE FACTORS

Actual make-or-buy decisions often involve subtle factors that can easily be overlooked. We must be sensitive to the need for both depth and breadth in our thinking when we are engaged in make-or-buy analysis. It is embarrassing to set forth a strong economic case for a make-or-buy decision and then have a member of top management knock down the whole argument by advancing a unquantifiable reason which we did not think of, could not quantify, or did not emphasize but which unquestionably carries decisive weight.

Such typical factors include:

1. Additional or lessened burden on supervision and higher management
2. Quality and delivery performance of vendors
3. Need for an assured source
4. Need to safeguard designs or process secrets
5. Aim of employment stability
6. Possession of needed skills
7. Raw-material availability
8. Foreseeable and suspected technological change
9. Effect on existing suppliers and their relationship to you

The Effect of Overhead on Make-or-Buy

The tremendous effect of how overheads, particularly manufacturing burden, are applied in calculating make costs is well demonstrated in the case problem. They can turn a good make into a seemingly obvious buy.

Writers on the subject agree on two principles:

In comparing a calculated make cost with a present buy cost, you should include in that make cost only the *additional* overhead that will be *newly* incurred if you decide to make.

In comparing a present make cost with a proposed buy cost, you should include in that make cost only the overhead that will be *eliminated* if you decide to buy.

These principles seem so clear that the frequency and depth of the resistance to them among companies and managers comes as a surprise. We have all been so deeply indoctrinated with conventional absorption-costing logic that we apply it when it is illogical.

Practical Points on Make-or-Buy

• As part of your make-or-buy program, use one or more standard formats, depending on the complexity of the situation. This gives you a checklist and some insurance against missing something. Some forms have been published, but it is better to develop your own, designed to meet your particular circumstances and needs.

• Remember that the relevant costs are *usually* the variable costs, not the fixed costs. The key word is "usually." A series of buy decisions can increase some of your fixed costs, such as supervision. On the other hand, buying parts previously made may not decrease your shop supervision nor decrease your other overhead. Be subtle in your thinking—but practical.

• In estimating costs for expected labor efficiency and projected scrap allowances, be conservative not optimistic. It may be safest to calculate efficiency at that of the least efficient supplier.

• Be cautious in using "average costs" in your calculations. Important parts represent a large percentage of total product cost and should be run most economically on the best machines. The result for such important parts should be better than average costs, and these should be the ones used in the calculation.

• On major make programs do not neglect your learning costs. They are inevitable.

• If you contemplate making, be sure you have access to a good source of raw material and at a price as good as that your present suppliers are paying.

• If you are going to make an assembly and its parts, remember that you will now be inventorying all the items, not only one bought item.

• Avoid the natural tendency to underestimate repair and maintenance costs. Be pessimistic and conservative.

• Have an independent hard-headed follow-up made on the make-or-buy decisions after they are in effect. Check that the savings projected for the make decisions are really being made and that the vendor costs for the buy decisions are not higher than originally projected. It is unusual to find this seriously done. In the case problem a follow-up would be made on actual hobbing times and scrap. For example, if gear A is not being hobbed in 3 minutes, you will not realize the projected $19,000 a year savings.

• Since all but the simplest make-or-buy decisions usually end up entailing a capital investment, make these decisions in the framework of a capital investment decision-making process. Screen them with all the others and measure their projected return on investment with a method that recognizes the time value of money.

• Be sure that Estimating and Purchasing, as well as Engineering, are keeping abreast of new technological developments. New processes and new materials can often enable you to make more efficiently something you had to buy in the past. Conversely, something you are now making might well be replaceable, at a good saving, with a vendor's new material or process. Technological change is a good reason for reviewing past make-or-buy decisions every 2 or 3 years.

• Do not be overawed by experts on make-or-buy decisions. Many of them conclude that it usually pays a company to continue buying what it has always bought and making what it has always made. This strong implication that there is little opportunity for profitable make-or-buy analysis may well not be the case in your company. You are looking for past decisions to make or buy that were originally made incorrectly or are no longer correct because of changed conditions and costs. Finding these mistakes and correcting them leads to reduced costs and a saving.

Using the Computer in Estimating

Chapter 9

For many reasons Estimating has lagged other management functions in using the computer, but the question is no longer: Does the computer have a use in Estimating? Now it is: How can we demonstrate that the computer can pay for itself in our particular Estimating situation?

Estimating's Lag in Using the Computer

In most companies, the first things on the computer have been payroll and sales statistics. Then frequently came inventory balances (receipts, issues, and balance on hand), followed by accounting records (accounts receivable, accounts payable, and the general ledger). It has not been unusual to see accounting on the computer before the inventory because usually the computer is part of the controller's domain. But if the books of account went on the computer first, the inventory records usually followed next.

Only after these four area are on the computer have other management areas such as Estimating, Production Planning and Control, and Purchasing had the opportunity to use the machine. In many companies today Estimating, as a management function, is still waiting to get on the computer, for a number of reasons.

194

• In many companies, Estimating has not had its body of practice, procedures, and steps clearly enough defined to permit ready use of the computer. In many estimating situations, large judgment inputs are involved, both in the product whose costs are being estimated and in the estimators themselves. The computer needs exactly defined steps and inputs to yield a meaningful output. In areas like sales statistics, payroll, and accounting the procedures and definitions are much more easily established.

• Many industries lack a body of estimating practice, and each company must pioneer its own use of the computer in estimating. Such pioneering tends to be expensive.

• Organizational realities have had their effect. In most companies since Marketing has great clout, their need for prompt and comprehensive sales statistics has tended to get first attention. Also, since in most companies the computer is under the Controller, accounting matters have a higher priority in using the computer.

• Some senior management has not been perceptive enough of the need for consistent, high-quality product-cost estimating. They have been willing to accept estimates less accurate or less complete than they could have been and those too long in the making. Such management does not give good product-cost estimates the priority they deserve and are loath to authorize expenditures for use of the computer for estimating.

• Finally, the importance of, and need for, good estimating follow-up has not been widely enough recognized. Without the data-handling capability of the computer, good estimating follow-up often cannot be achieved. Lacking management recognition, Estimating is denied the funds needed to use the computer for follow-up.

Signs show that this lag in Estimating's use of the computer is changing markedly. More and more of the estimators I meet are using the computer, at least to some limited extent. They are still in the minority, but their numbers have been increasing over the last 5 years. As would be expected, estimators who are using the computer tend to be working in large companies or in the larger divisions of large companies. Only rarer, more-progressive medium-sized and smaller companies are using the computer for estimating at this time.

Factors Favoring Estimating's Use of the Computer

In terms of the computer we are living in an almost revolutionary, not evolutionary, age. Most of us believed that the trend would be to larger and larger central main-frame computers, with all the various management functions on-line using the central computer and being serviced by the computer staff of systems designers and programmers. But it is not working out that way. The tremendous technological advances leading to the minicomputer and the microcomputer are drastically changing that trend.

In the years ahead, it is highly likely that Estimating, Purchasing, Production Planning and Control, etc., will each have its own minicomputer. These various areas will use a common data base and in larger companies may be tied into a central computer, but in many companies each management department will have its own computer and be relatively independent of the services of a central computer staff. For Estimating this will allow much greater flexibility and faster reaction time in responding to cost changes. It will also mean that estimators will have to know how to use the computer.

Here are some of the factors favoring Estimating's greater use of the computer:

DECREASING COST PER UNIT OF COMPUTER CAPABILITY

Costs have been declining sharply per unit of memory and per hundred lines of printing. The central processing unit (CPU) has greater capability and lower costs. This point is often made by showing an old photograph of an early room-sized computer beside a microprocessor chip under a magnifying glass, with the explanation that the chip has 250,000 times the computing capability of the computer in the photograph.

AVAILABILITY OF MORE AND LESS EXPENSIVE MODELS

This is particularly true in the minicomputer field. Today in fact, the problem for management is deciding which minicomputer to buy or lease from the many available.

MANAGEMENT'S INCREASING FAMILIARITY WITH THE COMPUTER

In the early years much money was wasted on computers, which were fashionable but both underused and misused. Managers were afraid of them. Today they are a way of life, and more and more managers and their staff feel easier in using computers. I find a certain wry humor in watching a production dispatcher go to a terminal, punch in a part number, and expect to see a listing of all the open shop orders for that part, the unfinished operations, the quantity in stock, and any reservations against that stock when 10 years ago, he would have been afraid to touch the machine.

THE TREND TOWARD DATA BASES

In estimating the computer will provide the mechanized integration of estimating data and Accounting's product standard costing.

THE GROWING REALIZATION OF THE IMPORTANCE OF ESTIMATING FOLLOW-UP

Accounting records provide the data for estimating follow-up. If Accounting is on the computer, obtaining follow-up data is much easier. In fact, the reason for too little estimating follow-up being done to date is the high cost of the clerical workload involved. When both Accounting and Estimating are using the computer, such follow-up is much more likely to be achieved.

To sum it all up, today every Estimating department can use the computer's capabilities. If you are already using the computer to some extent, the odds are that you could benefit from still greater use.

A good-sized investment of company money and management time is required for the start-up effort of using the computer for Estimating. The system has to be designed, tested, and field-use-debugged. A very strict requirement, after the installation phase, is updating the data used in the computer's estimating system. Prompt maintenance is vital if the computer is not to issue bad estimates.

Certainly no smart management should incur such design, installation, and maintenance costs unless there is reasonable assurance that prospective benefits will exceed the costs. The problem for the estimator is to determine how best to use the computer, how much benefit such use will yield, and how to demonstrate to senior management that the yield will return the cost involved. I would like to approach these questions from three viewpoints—time, type of industry, and specific estimating problems that may exist within the company.

Time Savings with the Computer

Perhaps the single best approach for the estimator in deciding whether a computer would be self-supporting is to consider it from the standpoint of *time*. Analyze the steps and the work load involved in developing an estimate in your specific estimating situation. Determine what step or steps take the most time. Then see how the computer can save you time on those steps. The following examples show how this will vary with the product and the type of industry.

Parts Listings

Estimating product cost when the product has many parts and four or even five levels of subassembly is a time-consuming chore. When some of the subassemblies in a new model have already been used on previous models, the estimator searches out the parts listings or bills of material for them. This means a lot of time spent searching through Engineering or Estimating files, or waiting while others do the searching. If the bills of material are in the computer, retrieval is fast and a lot of time can be saved.

Parts Routings

Some estimators are given parts sketches or drawings with the instruction "machine complete" or "machine as per blueprint." As a result, they have to detail the list of sequential operations required to make each part. If standard routings by type of part are on the computer, a lot of repetitious work can be avoided.

In metalworking today a service organization provides computerized parts routing and labor standards. From your list of machine tools and a parts print their computer develops the routing, the setup, and cycle time standards for each operation listed to make the part. Or you can be on-line to their computer for the needed routing and time standards development.

Labor Standards

Detailed estimating requires not only parts routing but the calculation of labor time standards for each operation on the routing. Time standards may be needed for both cycle (per piece) times and setups, where setups are required for the various operations. In turn, the work elements included in each operation have to be thought through and listed by the estimator. Finally, time values for each work element have to be developed. These steps involve a great deal of work. In some companies, estimators have this work done by Industrial Engineering or a Time Study department. Less fortunate estimators have to do it themselves. In either case, much time and work are required. In metal cutting, for example, setting an operational time standard means a lot of tables looked up, cutting-time formulas used, etc. If labor standards are computerized, the elemental times are easier to retrieve and operation and setup time values are more rapidly calculated. Today only a few companies have their labor time standards on the computer, but the number will inevitably increase.

Material Costs

When estimates are developed for large contract defense items, Estimating is usually required to use Purchasing's services to obtain, say, three bids for purchased components and at times even for raw material. Many estimators however, do not have the luxury of so much assistance and time in developing their estimates. Instead, they must personally estimate raw-material and purchased-component costs from files of supplier catalogs and price lists. Retrieving specific prices and costs from these sources can be a time-consuming job. If these material costs are computerized, both search and retrieval time will be drastically reduced.

All Calculations

I have deliberately saved for last the greatest time-saving potential of the computer—calculations. Most estimating situations require a great deal of calculation. And obviously on this matter the computer is a godsend.

• One estimator I know does estimating on large glass-lined steel tanks with rounded ends and many parts. To estimate a given job, with all the material-weight and labor-time calculations, used to take, on the average, 4 days. With a computer the input and calculating time is down to 20 minutes.

• In the electronics industry a continuing estimating chore is to develop time standards for printed-circuit-board assembly. Hand calculations took anywhere from 4 to 8 hours to calculate printed-circuit assembly-labor times, depending on the size of the board and the number and variety of the components involved. With computerized calculation even the most complex board can be estimated in 15 minutes.

• In costing a glass bottle, depending upon the costing method, there can be 13 multiplications, 3 divisions, and 5 subtotals required. If the calculation is computerized, you can cost 50 bottles in the time it takes you to do 5 bottles by hand.

As Leibniz, the mathematician, said, "It is unworthy of excellent men to lose hours like slaves in the labor of calculation."

Another point is worth making on the subject of computerizing the calculations required in developing your estimate. The more complex the product, the more calculations you must do, and the greater their complexity, the greater the chances of error. People make mistakes.

Machines like the computer do not make mistakes unless we program them wrong or they malfunction. Since errors in an estimate can be very costly, reduction of calculating errors is one of the chief attractions of the computer.

Thus, estimating time can be reduced by the computer in five areas: parts listings, parts routings, labor standards, material costs, and calculations. Not every estimator has potential time gains in all five, but many estimators have potential gains in at least two. It should not be difficult to convince senior management of the value of using the computer and even giving Estimating their own mini- or microcomputer.

Type of Industry

Another way of determining whether the computer has initial or further applicability in your own estimating situation is to consider the wide range of industries in which successful applications of the computer in estimating have been demonstrated. If other companies in your industry are benefiting from using the computer in estimating, why can't yours?

A very broad range of industries, covering a great many manufactured products, are already using the computer in product-cost estimating. These examples are by type of industry.

TYPE 1

The company makes only one or two major products, but each product has hundreds or thousands of different models or types.

• A company makes only electrical or electrical-mechanical controls, but it makes thousands of different kinds of controls.

• A glass-bottle company makes only glass bottles, but it makes hundreds of different types of bottles of varying weight, configuration, and color.

With such conditions, a computer makes it easy to turn out prompt and consistent estimates of product costs. It can also print out and keep in computer records the estimates already made for the hundreds of product types. The output can be used as input to product standard costing. Finally, the computer can provide a mechanized follow-up of actual costs for individual Estimates, something you might not be able to afford to do manually.

TYPE 2

The company custom builds products. Each order is for a unique variant of the basic product. Many orders are one-offs, and the rest are for very limited quantities. Examples are a heavy-equipment manufacturer or an industrial builder.

A company of this type must worry about the completeness of its estimates. Its products have many parts, both manufactured and purchased, which must all be included if the estimate is to be accurate under the conditions of the great differences between orders. The computer can help assure estimate completeness, facilitate job actual costing, and provide mechanized follow-up on actual performance to estimate which is so essential in this type of business.

TYPE 3

The company is job manufacturer, producing a product to order, with a minimum of the items being of the stock or proprietary type. Normally, there are numerous requests for quotations. Competitive pressures demand prompt response and tight cost estimating. If orders result from the quotation, they may be in the hundreds, thousands, or tens of thousands, though the value per unit may be quite small. In such companies the orders received are often only a small fraction of all the quotations made, but it is wise to respond to all requests for quotations to build up (or at least maintain) the company's sales volume. Examples are a job machine shop, a paper-box company, and a printing house.

Many of the computer applications in estimating have been made in this type of company. The need for the company to achieve prompt, consistent estimating on a great number and variety of quotations is obvious and may even become acute. The great data-handling capability of the computer can meet this need.

Examples of Specific Computer Applications

A JOB MACHINE SHOP

A large job machine shop uses a batch approach on its computer to develop estimates for over 400 requests per month for quotations. The system applied to each job encompasses operational routing, machine selection, and then cost estimating. The sequence is as follows.

Step 1

Input is a complete parts description with detailed specifications describing the part and its characteristics. The means of input are a series of input sheets covering identification information, material and order quantity specs, end specs, diameter specs, hole specs, etc. From these input sheets, punched cards are produced.

Step 2

Input errors are diagnosed by a series of subroutine subprograms to detect inconsistencies, errors, and omissions. Error messages are printed out and serious errors terminate the program, with the cause shown on the printout. Smaller errors or omissions merely cause the computer to insert assumed values, but these and the adjusted-for errors are noted on the final printout by the computer. The input subroutines also calculate finished and stock weight of the part being estimated.

Step 3

Five subroutines, constituting the process-routing segment of the system, select the operational routing required to produce the part. The operations are listed in the output printout with an operation number, operational word description, detailed machining information, and a list of tooling required for each operation.

Step 4

Fifteen subroutines then select the machines to be used for each operation, depending on such product specs as size, tolerances, and finish requirements.

Step 5

The next series of subroutines generate the standard time needed for setup and per piece cycle for each operation and total time per piece for the specified order quantity (setup time divided by order quantity plus cycle time).

Step 6

Then costing rates for each type of machine are applied to the cost per piece. Costs for material and outside operations are calculated from accessed file of purchased costs. The company uses direct costing, and the marginal contribution of each quotation as estimated is calculated.

The system is designed to generate a quotation for up to six different order quantities for any given quotation. All the important input data are shown as part of the output printout. The estimate is manually reviewed by Estimating personnel, but indications are that their work has been reduced 70 percent over the manual estimating previously done. If an order is received from the request for quotation, the estimate printout becomes the basis for the subsequent shop order.

CERAMIC ROD AND TUBING MANUFACTURER

A company making ceramic rod and tubing uses its time-sharing computer system to develop product-cost estimates for an infinite variety of product sizes produced from several different ceramic materials. The estimator punches a tape, enters the computer via a telephone hookup, feeds the tape into the computer through the tape-typewriter consoles, and receives the estimate as a typed printout on the console.

For ceramic tubing the input data are the inside and outside diameters, length, tolerances, material, chamfer, and/or radius dimensions with tolerances and quantity required by the request for quotation. The computer calculates blank weight, selects the necessary type of forming machine to use and the decimal hours per piece for forming for that weight of piece. If grinding is needed for tolerances, chamfering, or radius forming, the decimal hours per piece for cost of grinding are calculated. The system determines the instances and cost of the sequential inspection operations needed. Costing for each type of operation is done by dividing the departmental cost per hour by pieces per hour at each operation.

The system prints out a complete operation sheet listing each operation. For each operation it lists production rate, setup and cycle cost, and the total cost per operation for the quantity being quoted. This is then adjusted upward for scrap loss, the loss additive varying directly with the quality of material specified. That is, the higher the material quality, the higher the scrap-loss additive. With a markup factor added, the estimate becomes a selling price on a quote sheet.

AN INDUSTRIAL BUILDER

An industrial builder uses a computer to estimate the detailed costs of a building for a given length, width, and height. With more detailed input, the computer system gives a complete price breakdown by type of material and type of labor skill required for the five trades employed.

The system design includes a reasonability test using 20 key ratios to detect omissions in the estimate. The ratio tests were developed from actual costs on past jobs. By comparing the ratios in the new estimate with these "standard" ratios, important variances are printed out to indicate possible omissions or areas of overbid.

The system calculates not only worker-hour distribution by type of trade but also standard materials and standard equipment required. Computer terminals at job sites can be used if desired.

Project engineers use the computer estimating system to compare and choose between alternatives in such important variables as architectural styles, building size, and heating and ventilating systems.

A GLASS-BOTTLE COMPANY

A glass-bottle company uses its computer to develop product-cost estimates for each of the hundreds of different bottles it produces. Many are packed in different types of cartons, and cost for each of these packs is estimated. The company does not yet have on-line capability and uses a discrete run or batch approach.

Inputs are bottle weight, glass color, bottles per carton, and type of carton. Each Production department has its own variable or direct costing rate and an overhead rate applied per dollar of direct cost. All estimates are calculated by the computer in terms of cost per hundred gross. Bottle color determines batch material composition and cost per hundredweight. Bottle weight determines batch hundredweight per hundred gross, melting costs, and forming speeds and costs. Bottles per carton and type of carton determine carton and carton-forming costs and select and pack costs.

Using the computer has markedly improved the company's ability to estimate new bottle costs promptly and to make consistent estimates.

METAL-STAMPING COMPANY

The company produces high-quality metal stampings of a very wide variety for luggage, trunks, plumbing specialties, appliance hardware, cabinets, control assemblies, etc. Sales volume is under 2 million a year, and the company employs approximately 60 people.

The following problems led to the computerizing of estimating:

Incomplete or inaccurate estimates and even guesswork that adversely affected profitability

Too large an estimating backlog and loss of business because of estimating delays

No costing listing and thus no estimating follow-up

Estimates that were inconsistent or did not properly reckon in such factors as lot-size variables

Cost factors stored in the computer are material, setup times, production time, labor rates, and such miscellaneous costs as heat treating, plating, packing, and purchased components. The costs needed for a given quotation are extracted from storage and entered directly into a data base for processing. Total job costs are combined with the desired markup. Any applicable quantity discounts are applied to develop a bid or selling price. Costs and selling prices can be calculated, of course, for any ordered quantity.

PATTERN SHOP

This is a $2 million a year business in which the job cost estimate is still developed manually but the computer is used for estimating follow-up. Direct data-entry units on the shop floor are used to record start and finish times. Material costs are also entered. Thus labor and material costs are collected as incurred for each job. Periodic reports show actual cost and state of completion against estimated costs and delivery promises. A special flag indicates each job on which actual costs to date have reached 80 percent of quoted costs to allow time for corrective action.

These are only six examples of specific applications of the computer in the estimating in widely different industries. Many more are available for airframes, engines, electronics, soft goods, box making, printing, etc.

The Computer's Contribution to Specific Estimating Problems

The third approach estimators can take to determine the computer's applicability in their own situation is to consider what specific estimating problems the company is currently facing. Two not uncommon estimating problems come to mind.

LACK OF ESTIMATING CONSISTENCY

A lack of consistency in estimates means that some orders are overbid and others underbid. If the problem is one of unclear and incomplete specification, the computer *cannot* help, but lack of consistency may also be observed when different estimators in the company develop estimates for different but similar jobs. The same estimator at different times on similar jobs may see the jobs differently and thus develop disparate quotations.

Obviously the computer offers a viable solution. The machine can only estimate as it has been programmed to do. If it is programmed correctly, if the cost data it is fed are up to date, and if the job specs are complete, the estimate it develops will be correct and consistent. With cost data that are not up to date, the estimates will be consistent but not correct.

OVERWHELMING ESTIMATING LOAD

When the estimating workload is too heavy for the available worker hours, the promptness with which requests for quotations are handled leaves much to be desired. The backlog of requests for quotations is high and stays high. As a result Marketing management is hampered, spending too much time expediting requests for quotations with less time for selling. This can affect company volume now and in the future. Sometimes this condition includes careless and inconsistent estimating because of the time and workload pressures.

With its great computational speed and its incredible data-handling capability, the computer can produce a lot of estimates in a short time. If the conditions outlined above are causing the loss of profitable jobs and/or slowing the growth of the enterprise, the computer may be more than self-supporting.

Twelve Cautions about the Computer

Proper use of the computer in estimating demands time of Estimating management, which must design, install, debug, and maintain the system. The advantages of the computer are obtained only with a time and cost investment. As managers we should have a hard-headed attitude toward the computer. Such an attitude and the actual experience of others requires us to be sensitive to certain dangers that may be critical to the cost-benefit success of computer use.

1. RESEARCH THE FIELD

In developing and designing plans for using the computer in Estimating it is wasteful and senseless not to use the experience of others. Trade or industry associations can tell you what other companies in your industry have done. Then contact those companies. From them and from people you know in the industry find out how others are using the computer for Estimating. You will be surprised how much you can learn. In some industries where geographical distances and freight costs restrict the competitive areas other companies outside your area will usually be quite willing to demonstrate their installation. Successful users generally are proud of their accomplishments. Research the literature, particularly the various trade and professional journals. A wealth of real and practical experience has already been published. If possible, try to visit some of these companies for demonstrations. Don't reinvent the wheel.

The two barriers to such research are shyness and time. You may feel that it is an imposition to ask an Estimating manager in another company to give you time from a busy day to show you his system. Don't. You will be amazed at how willing managers will be to discuss their successes and the problems they encountered. Despite an occasional rebuff most professionals like to discuss their trade with other professionals, and you in turn will have opportunities to help others.

Such research takes time, but it is a good investment. Every hour you spend on such research can later save you many more hours of misdirected effort and error correction that others have already spent. Also, you are not all-wise. You will not think of everything. By seeing what others are doing and have accomplished, you can pick up ideas you might not have thought of. Research time can well yield both a better system and avoid misdirection and error.

2. SHOP THE COMPUTER COMPANIES

If you are going to be using your company's present computer, your flexibility is limited. But it is still wise to see what other computer companies have available and have installed in the estimating area. All computer companies have manufacturing specialists who assist the salespeople. Talk to them, particularly those working for the manufacturer of your present computer. Talk to more than one company so that you have a grasp of what is available and already working elsewhere. Such shopping can open doors to demonstrations that have parallels to your own estimating situation. This shopping is particularly important if you are going to have your own minicomputer in Estimating. The variety available is overwhelming.

3. BE CAUTIOUS OF SOFTWARE PACKAGES

Apply great caution in considering software packages. Estimating is a very individualistic management function, often with wide differences between companies in the same industry. Since equipment, operating conditions, and cost accounting procedures can all vary between companies, software package used successfully elsewhere may be inadequate or require expensive adaptation. In any event, a detailed demonstration on-site of an actual installation, plus your own careful study, is essential before buying any package.

I have never met any manager whose company bought a software package, installed it as is, and was happy. I know many who bought a software package, customized it to their situation, and were pleased. It seems universally true that you must custom-fit any software package.

At present software packages for estimating use are relatively rare. In some industries there still are none, but this will change. Wise Estimating managers should keep alert to developments in their industry, because as software packages become available they are worth investigating. If they can be customized to fit your individual situation, a great deal of time and money can be saved.

This is a fast-changing area, and potentially valuable developments are coming. For example, in the Pacific Northwest, there already is a computer service firm that offers an integrated job-shop system. The firm supplies all the hardware, the computer software, the installation help, and equipment maintenance service. You buy the complete package. Their system includes the following hardware from four different manufacturers:

One 4/30 CPU with 64K 700-ns memory and eight channels of I/O

One 10-million-byte disc drive

Two video terminals

One 300-lines per minute printer

The following software application packages are included in the deal:

1. Estimating
2. Accounting
   a. General ledger
   b. Budgeting
   c. Fixed asset accounting
   d. Accounts payable
   e. Payroll
   f. Job costing
3. Purchasing
   a. Vendor information
   b. Purchase order
   c. Inventory
   d. Accounts payable
4. Marketing
   a. Customer Information

The package costs include 280 hours of their staff's time for installation, training, and follow-up support.

The package can be purchased for $55,000 (as of 1980) or leased at approximately $1100 per month. Equipment maintenance costs are $310 per month. An expanded system is offered with 30 million bytes and four video terminals for $60,000 and $385 per month maintenance costs.

The beauty of such a system for a small and medium-sized job shop is that a total package is available, the equipment, the software, installation help, and equipment maintenance. In the future, such a computer service idea will be available across the country. I know of two similar ones on the East Coast. It would be worth investigating by job shops in such industries as printing, machining, punch press, die casting, plastics moulding, wood working, box making, and so forth.

4. CLEAN UP YOUR PRESENT SYSTEM

Before computerizing your estimating, correct deficiencies of your present manual or semicomputerized estimating procedures that would be carried into the new system. The new computerized system will not compensate for the deficiencies but only preserve and solidify them. It may even exacerbate their effects. For example, if you are estimating with incomplete or inaccurate product specifications, incorrect parts routings, erroneous cost-variance data, inadequate estimating formulas, bad overhead allocations, bad past actual costs because of bad shop counts, etc., do not go near the computer until you have cleaned these problems up. The GIGO effect (garbage in, garbage out) may be a cliché, but it is still true. You should only go into the computer with a good manual estimating format and reasonably good estimating product and cost data.

5. ESTIMATING'S RESPONSIBILITY FOR THE SYSTEM'S DESIGN

When Estimators first begin using the computer and working with computer professionals, jargon like bits, bytes, RAM, and ROM may frighten the estimator into thinking "Oh boy! This is the computer. I'm Estimating. I'll leave it up to them." Right then and there the troubles start.

Only Estimating management can be responsible for the design of the computerized estimating system. It must specify the input data to be used, the records to be kept in computer storage, and the output data needed, because estimating is their responsibility. Such specifications cannot be delegated to the computer systems designers. They know a lot about computers, but they cannot be expected to know what Estimating management knows about estimating. Their inputs should be solicited and listened to, but the responsibility for the final design must be that of Estimating management. Otherwise the system as designed may well be less than what you want and need. Once designed and installed, it is difficult to change.

6. TIE-IN WITH OTHER MANAGEMENT AREAS

All management areas are tied in with each other. Consider how Estimating is affected by Design Engineering, Production, Cost Accounting, Industrial Engineering, etc. They all affect you, and you affect them.

Design your estimating system, wherever possible, to tie in with other management functions. This is thinking in terms of a data base, in which the output of other areas is part of your input data and some of your output data are input for other areas. For example, you may be able to use present computerized engineering bills of Material, and your estimating output will be valuable input to the company's standard costing and cost accounting.

Tie-in has value in another context. When some of your output data and reports can be of great value to other management areas, often a slight change in output format, without affecting Estimating in the slightest, can make these output data of even greater value. As you start to complete your designs, show them to these other managers. Too often at management meetings when a computer report is before the group, you hear someone mutter, "Oh, I didn't know we had these data."

7. DESIGN FOLLOW-UP INTO YOUR SYSTEM

Never lose sight of the importance of Estimating follow-up unless you want to be condemned to repeat the errors of the past. When you design your computerized Estimating system, build in the ability to collect and compare eventual actual costs with estimates. It may increase the design costs and the later computer-use costs, but the benefits should more than pay for the added expense.

In many Estimating situations, the effective follow-up data have to be collected by Accounting and Estimating must look to Accounting's computerized data for the needed follow-up facts. If Accounting is already on the computer, as they so often are, Estimating's computer system design must be such that accounting data are compatible and thus available for follow-up use.

8. FULLY DOCUMENT YOUR SYSTEM

A statement all too often true is: *No major computer system is ever installed on time, within budget, and with the same staff that started it.* That last phrase is particularly pertinent. Often changes wanted on an existing computer output report are too expensive or even impossible to effect because the original system design was inadequately documented.

Be sure the system as it is being programmed is also fully documented. No matter how good you are and how good your systems designers are, you will not design the ultimately perfect system. A year after the design is in and working, you will have ideas for its further improvement. Or operating conditions will have changed requiring changes to the design. Whatever the reason, many of these changes and refinements can readily be made if the system installation was well documented. If not, such changes are expensive, sometimes even requiring a new system design and program. Many a computerized system exists as it does today, inadequate and ripe for improvement, because it was not properly documented and thus is too expensive to update or correct.

9. PARALLEL YOUR NEW SYSTEM

Never let anyone persuade you into dropping the old system when you start the new one. This is bravery with your money, not theirs.

When your new computerized estimating system is ready for use, parallel it for a time until it has been field-use-debugged. Some startling surprises can develop when the system is used on real jobs under real conditions. If you are paralleling with your old procedures, it is possible to continue estimating. It is more costly than not paralleling, but it is good insurance. How long to parallel will depend on the situation. Usually 1 month is too short; for most situations, 3 to 6 months will be adequate. Just be sure to establish a cutoff date for the parallel work.

10. DESIGN FOR EASY UPDATING OF DATA

Of all the cautions being reviewed this is probably the most important one for Estimating. Once the computerized estimating system has been installed, the data must be kept up-to-date. Material and labor costs change, new equipment with new production rates is installed, and other aspects of the operation change. If the cost effects of these changes are not continuously inserted into the computer, the machine will simply crank out increasingly poor estimates. This represents one of the most serious and prevalent dangers of the computer's use in estimating. In one installation, for example, the computer was using costs for equipment that had long since been replaced by newer and faster equipment. When you plan on the use of the computer in estimating, include procedures for prompt revision of the cost and product data in the computer. And remember, the easier the change procedures are designed for the user, the more likely that they will be used.

11. USE EXCEPTION REPORTS WHEREVER POSSIBLE

Wherever you can avoid those thick computer printouts which no one ever has the time to read and use. Modern printers are so fast that their uncontrolled output can inundate the user. Many computer people are insensitive to this problem. One Sales vice president I knew was given monthly sales statistics over 6 inches thick. Not many sales vice presidents will plow through that much information.

With estimating follow-up data, for example, if you have 200 orders a month or more completed, you will be interested in operational cost detail when costs are *widely* less or greater than estimate. If actual costs are within acceptable limits on many jobs, a total estimated versus total actual costs by job for those jobs may well be all you have any use for. Exception reporting requires you to establish variance standards and ranges within which you do not want detail and beyond which you want detail to be printed out.

12. DESIGN REPORTS FOR READABILITY

Most computer reports are too repetitious and crowded with data. As one estimator put it, "Our computer people seem to feel that if there are 132 fields across the report, they are not doing the job if they don't print in every one."

It is axiomatic that the simpler a report is and the easier it is to read, the more likely it is to be read and used. When you design your output reports, put yourself in the chair of users and ask yourself what they need to know and how they are going to use the data. Avoid repetitions of part numbers on successive operations. If the part number applies to the first operation, it obviously will be the same part number for subsequent operations on the same part. Look at some of your present computer output reports with a critical eye. Can you redesign them to eliminate repetitious information, increase their white areas, and improve readability?

Unless you are sensitive to this feature, you will not have the most readable reports. The key word to the programmer is "supress." Since that means more programming work, it is not likely to be done unless you insist on it. But it is more work for the programmer only once, and it will simplify your work for as long as anyone uses the report.

Estimating Capital Investment Costs

Although this book is concerned primarily with product-cost estimating, this chapter deals with estimating capital investment costs for two reasons: (1) many product-cost estimators are involved with estimating capital investment costs, and (2) one of the four steps of capital investment decision making is all too frequently marked with repeated and serious errors, which the talents and resources of product-cost estimators can help reduce.

The greatest detriment to steadily improving product-cost estimates is the time strictures imposed on estimators by their current workload. Since most estimators already have too little time to develop ways of improving their results, deep involvement in capital investment cost estimating reduces their time for improvement even more. Nevertheless, the fact remains that they are involved in capital investment in many companies.

Four Steps of Capital Investment

The four steps of capital investment are:

1. Estimating costs of the investment and (if savings are involved) the projected savings that will result from the investment

213

2. Calculating the return on the capital investment

3. Selecting the proper capital investment

4. Making the postaudit

Unfortunately, most of the attention in writings on management is paid to discussing the second step, how to calculate the rate of return on a proposed capital investment. But surveys of managers show that the greatest and most recurring problems are in the other three steps. This is an important point. We in management must be continually aware of it and spend more time on steps 1, 3, and 4 of capital investment decision making.

Estimating the Investment's Costs and Savings

Some capital investment projects do not involve savings. Projects on the buildings and grounds, for example, can entail capital investments, but normally savings, per se, are not involved. This discussion is concerned with capital projects and investments that are expected to result in savings.

The common experience in this step is that the most persistent errors are underestimation of investment costs and overestimation of projected savings. Thus this step becomes the most important and potentially most fruitful area of improvement in capital investment decision making.

The reasons for these errors are readily understandable. Managers want their share of the capital investment monies, and the brighter the picture they paint, the more chance their project has of being approved. For senior management these errors mean that the capital budget is not expended as effectively as it should be. In fact, an awful lot of money can be wasted. It is sad to see, as I have, a quarter-million-dollar piece of equipment that never worked properly rusting beside the plant wall or an overhead parts-conveyor system used only on the one day of the year the board of directors makes a plant tour.

The areas of under- or overestimations are many. Readers can develop a list of their own from their own experience. Here are a few from my own observation:

1. On construction projects
   a. Failure to recognize and to state facility needs in the original design, resulting in extra costs for changes during and after construction
   b. An incomplete punch list, leading to failure to get a complete job from contractors
   c. Costly time delays because of inadequate project planning and/or weak follow-up

    d. Failure to include additional maintenance labor capacity and maintenance parts and supplies for the new building and the additional square footage of plant or office space

2. On equipment and tooling

    a. Not reckoning in your own company's costs for your engineering time, maintenance hours for installation and troubleshooting, additional power, air, or water lines, etc.

    b. Failing to include auxiliary attachments and/or accessories needed for full use of the new equipment

    c. Buying undersized units and having to upgrade

    d. Not including the cost of additional steps in prior or later operations required for proper use of new equipment

    e. Not recognizing higher purchasing costs for material needed by new equipment or tooling

    f. Failure to include higher costs incurred for labor during learning or break-in period

    g. Overoptimism about expected feeds, speeds, and production rates to be obtained (usually this is the result of too ready acceptance of the equipment salesperson's promises; always ask to visit a previous purchaser's plant, talk to the operators and the supervisors using the equipment, independently of the salesperson)

    h. Failure to include additional maintenance equipment, supplies, and spare parts

    i. Understating the outside service maintenance costs paid to the manufacturer or its local agent

I believe that this overoptimism, resulting in understated investment costs and overstated projected savings, comes from improper mental orientation. Senior management should take a stronger approach, inculcating and rewarding the proper attitude to this estimating task. I can add a happy example.

The consulting assignment was to lay down a 5-year plant layout program for a silverware manufacturer. The first phase was to reincorporate into the home plant a branch producing stainless-steel flatware. The actual moves and relocation of operating departments and stockrooms were to be made by the Plant Maintenance personnel under the company's plant engineer, a Scot getting close to retirement, who was telling me just how many maintenance labor hours he would need for all the relocating steps involved in the layout program. As I was totaling it all up and preparing to write the final report, I suggested to the vice president supervising the entire study, "Walter, let's add 10 percent to Scotty's figures for insurance." It seemed to me a conservative but wise allowance. His answer was, "No, Larry. I know what you mean, but not with this guy. He has taken great personal pride, as long as I've known him, in always bringing in his jobs on target. If there is any safety factor, he's already included it."

I thought then, as I do today, that that management was well served by that man. When they made a capital investment based on his projected costs, *they knew their downside risk.* In contrast, many senior managers read a capital investment proposal and mentally add some percentage to the estimated costs as they ponder their decision. Many plant or manufacturing engineering groups consistently have actual project costs that are greater than their original estimates, and the pattern is allowed to continue. If senior management continues to permit it, why not? Can't this orientation be changed? Why not stress the need for better project-cost estimates and then specially reward the engineers who complete their projects within their estimate? Wouldn't such special rewards inspire the other engineers to adopt the same approach? If more managements had a better knowledge of their downside risks in this capital investment area, there would be more effective use of the company capital budget and investment monies.

Life-Cycle Costing of the Investment

Theoretically, when the estimated costs of a proposed capital investment for equipment are being developed, the equipment's maintenance costs are included. Thus the cost of spare parts and of at least preventive-maintenance labor are projected and included. In practice this is rarely done. The commonest procedure is to estimate the equipment's acquisition cost, installation cost, and, if applicable, tooling cost. It is rare for attempts to be made to quantify and include the cost of spare parts and of preventive maintenance over the equipment's projected life. Instead these cost factors are treated judgmentally on the basis of the manufacturer's and the product's reputation for "reliability."

A more comprehensive and sophisticated approach is to estimate the life-cycle costs of the proposed investment. For a piece of capital equipment not only the equipment's acquisition, installation, and tooling costs but also the cost of spare parts and preventive maintenance work over the equipment's life would be included. As one financial executive of an English company pointed out, such costs may be two to three times the acquisition cost.

In estimating life-cycle costs of a capital investment, a starting point can be the manufacturer's suggested spares kit and proposed preventive-maintenance program. Frequently such maintenance steps may be named without the appropriate maintenance labor hours required, and the capital investment estimator must develop and include these labor costs. Manufacturer's data can only be considered a starting point and usually, to be safe, must be factored upward to develop realistic life-cycle costs.

There is a direct parallel between capital investment life-cycle costing and the life-cycle costing mentioned in Chapter 5. In defense products it is increasingly a requirement. In the future, it will probably become more common in the capital investment area for production and service equipment in nondefense situations.

Product-Cost Estimating's Role in Capital Investment

In many companies the estimates of an investment's costs and projected savings are made by plant engineers, production or manufacturing engineers, or industrial engineers, the people who should do such estimating. In smaller companies, where the product cost estimator has to wear several hats, that person frequently is also involved in capital investment cost and saving estimating. I do not like this situation because it reduces the time the estimator has available to improve the product-cost estimating performance, but frequently there is no alternative.

Let's consider the larger organization, where product-cost Estimating normally does not make the estimates of capital investment costs and savings. In this situation on this first step of capital investing product-cost Estimating can render a valuable service by *reviewing* the estimates of capital investment costs and savings. Product-cost estimators are necessarily hard-headed realists. They have to be if they are to make realistic predictions of what it will cost to make the product. In their review and study of the project's estimates they may well detect areas of possible additional costs the engineers did not think of and/or overoptimistic savings projections.

This is not a popular role I am proposing, and it impinges on the time available for their basic and most important job. Perhaps additional product-cost Estimating personnel will be needed. But the problems in underestimating capital investing cost and overestimating savings projections are so common and expensive that this is one way the necessary talent can be brought to bear on such problems.

Calculating Return on Capital Investment

The second step of capital investment is determining the rates of return of the many possible investments management is being asked to make. Since the capital budget is never large enough to make all the promising investments, we must decide which will give us the best returns.

To review this rate-of-return concept, let's look at the operating results of two grain terminals:

	Terminal A	Terminal B
Sales	$4,000,000	$3,000,000
Profit	$1,000,000	$ 450,000
Profit margin = profit/sales	25%	15%

Terminal A looks much better than terminal B, but let's look a little deeper.

Investment	Terminal A	Terminal B
Buildings (20 yr)	$1,000,000	Leased
Machinery (10 yr)	3,000,000	$100,000†
Total investment	$4,000,000	$100,000
Investment turnover = sales/investment	1	30
Return on investment = profit/investment or		
(profit margin)(investment turnover)	25%	450%

† Improvements.

Now terminal B looks like a real deal.

This extreme example happens to be an actual one and shows dramatically that this return-on-investment concept is needed in evaluating investments.

To get a little closer to the situation most of us experience, let's compare two other situations:

	Operation change 1	Operation change 2
Product sales	$1,000,000	$1,000,000
Profit	$100,000	$100,000
Profit margin	10%	10%
Effect of change on profit	+$50,000	+$25,000
New profit margin	15%	12½%

It looks as though change 1 would be the best and should be done first, but let's look at the capital requirements.

	Operation change 1	Operation change 2
Investment to make the change	$50,000	$10,000
Return on investment	100%	250%

Now change 2 would be the first to make. Obviously, rate of return *must* be a factor, if not the most important one, in evaluating capital investments.

In the extensive literature on computation of the rates of return on capital investment the experts occasionally generate a good deal of disagreement. The overwhelming majority of companies use one or more of the following five methods to calculate rate of return of proposed capital investments. Although the first two are the simplest, they suffer from a basic and important defect.

PAYBACK METHOD

This method calculates the number of years of project life it will take to recover the initial investment. Popular variations of this method are:

Original net investment divided by initial annual earnings

Original net investment divided by initial annual earnings after taxes

Original net investment divided by initial annual earnings after accelerated depreciation and taxes

Original net investment divided by initial annual earnings after straight-line depreciation and taxes

A variation of the payback method is the application of a given interest rate on the capital invested. The effect is to lengthen the payback period considerably.

The payback method continues to be the single most widely used method for evaluating return on investment. It is the most easily understood. At estimating seminars I used to joke about its being popular because it is the only method really understood by the chairman of the board. But a participant pointed out a practical reason for its popularity. His company is in an industry (electronics) with a high rate of technological change. When his company has to choose between two investments with a like discounted cash flow rate of return (see below), they properly choose the one with the shortest payback period because the dangers of technological obsolescence are less with that investment.

Acceptable payback goals vary markedly between companies. A few companies have the policy that any payback period greater than 1 year is unacceptable. This seems too harsh, too high, a standard. From seminar questions it appears that most companies look favorably on a 2-year payback and use that as their goal. A somewhat smaller percentage accept a 3-year payback. Very few appear to accept more than that. An exception is the investment in a brand new product and/or plant, in which cases expected payback periods of 6, 7, 8, or even more years are common.

ACCOUNTING METHOD (FINANCIAL STATEMENT METHOD)

This method calculates return on investment by dividing project net saving (income) by the investment in the project. Arithmetically, it is the reverse of the payback formula. Variations of the formula are:

1. First year
   a. Initial annual saving divided by original net investment, i.e., new machine less salvage value of old
   b. Initial annual saving after taxes divided by original net investment
   c. Initial annual savings after depreciation and taxes divided by original net investment
2. Life of project
   a. Average annual net saving divided by average book value of investment (average investment is midlife value under straight-line depreciation)
   b. Average annual net saving after taxes divided by average book value of investment
   c. Average annual net saving after depreciation and taxes divided by average book value of investment

Though simple and therefore attractive, these first two methods do not take into account the *time value of money*. They do not recognize that a return received next year is more valuable than an equal return received 2 years hence. For example, an investment with a return of $10,000 per year for each of the next 5 years is not as good as an equal investment with a return of $25,000 per year for each of the next 2 years. In the second investment, you receive your $50,000 total return sooner.

For this reason, these first two methods are not really measures of the *rate* of return, and discounted cash flow (DCF) and present worth or present value (PV) are generally accepted today as the proper methods to use. Every major corporation uses them today, but a surprisingly high proportion of medium-sized and small companies have yet to adopt them.

DISCOUNTED CASH FLOW (DCF) (INVESTOR'S METHOD)

The discounted rate of return on an investment is that rate which makes the cash earnings over the life of the investment, when discounted, just equal to the initial investment. In effect, it is the interest percentage which, when compounded on the projected yearly savings from the investment, will yield an amount equal to the initial investment by the end of the investment's estimated life. The higher the interest earning indicated, the more attractive the investment.

This method entails the following steps:

1. Establish for each year of the projected life of the investment
   a. The expected investment (normally first year only)
   b. Cash inflow
   c. Net cash inflow
2. Obtain from compound interest tables the necessary multipliers for the years of indicated investment life, at successively higher rates of compound interest in, for example, 5 percent steps
3. Apply the successive yearly multipliers to the net cash inflow of successive years and total the results over the projected investment's life
4. Interpolate between those two compound interest rates whose totals bracket (one under, one over) the amount of initial investment to estimate the final rate of interest

Steps 2 to 4 are saved by using a computer or hand-held calculator programmed for the DCF calculation.

PRESENT-WORTH (PRESENT-VALUE) METHOD

This method is similar to the discounted cash flow method in that an acceptable or required rate of return is assigned to the projected net cash flows over the projected life of the investment to determine the projected present worth of the expected cash inflows. If these expected discounted cash inflows for the life of the equipment exceed the investment required, the projected investment will yield a rate of return greater than that required by the company and can then be considered as an attractive investment.

IMPROVED MACHINERY AND ALLIED PRODUCTS INSTITUTE METHOD

This method attempts to determine how much saving there would be in the next operating year with the proposed investment. It develops a relative return which becomes a ranking factor, or *urgency rating*.

Five elements enter the relative return determination:

1. Net investment
2. Next year's operating saving
3. Next year's use of capital that will be avoided because this year's capital disposal value is greater than next year's
4. Next year's use of capital incurred because of decline in the value of the proposed capital equipment in the next year
5. Next year's adjustment to income tax

Although this MAPI approach is frequently mentioned in the literature, it is not widely used.

I have purposely avoided details and examples of how these various return on investment measures are calculated. The two National Association of Accountants Research Studies listed in the Bibliography are inexpensive and give clear expositions with examples. If you are involved in capital investments, these two guides should be in your personal library.

More attention seems to be paid to calculating the return on capital investment than to the other steps combined. But the results of the calculations are meaningless if the inputs of the investment's costs and projected savings are not reasonably correct, if the needed follow-up or postaudit is not done, or if management does not take the action to see that serious errors are not repeated in the future. The fact that these matters are continuing problems in so many companies leads me to believe that calculating return on investment is much less important than estimating the investment's costs and savings.

Selecting the Proper Capital Investments

Some capital investments do not need selection. They just have to be done. If the plant needs a new roof, the money has to be invested, or, if ecology laws require plant modification, there may be little choice. But the common situation requires management to choose between many requests for capital investment.

Every large company today has a capital budget. Actually every company has a capital budget because even in the smallest company the owner has to have some idea of how much the company can afford to spend for capital investment, i.e., new fixed, depreciable assets. In this third step of capital investment, the problem for senior management is what choices to make from among the many offered.

Theoretically, those investments with the highest rates of return will be selected until the capital investment budget is expended. But, as you and I know, that is not always the case. The dynamics of personalities, management ego, existing political structures within the organization, most immediate past profit performances, and many other factors cause deviations from what theoretically should be done.

What has always concerned me in this step is whether everyone is getting an equal ear from senior management. For example, in some situations, the division manager with the strongest political connections at the corporate level may tend to get an undue proportion of the capital budget. Or the divisions with the highest profit and/or greatest return on investment may get the lion's share of the capital funds available, while other divisions with lower returns get much less, even though additional capital investment would enable them to improve their performance. How can we make it more likely that the capital investment selection will be reasonably dispassionate? How can we ensure that everyone will receive an equal hearing?

Another problem is that the selection of capital programs offered to management for approval is often the result of a haphazard collection of divisional or departmental projects which various managers would like approved. Senior management may not be presented with a group of projects resulting from a cooperative, organized search for investment opportunities which will benefit the entire company over the longest possible time. This is frequently management's own fault because they rarely assign specific responsibilities for generating such a systematic search for opportunities.

A possible solution for both these problems is a capital investment committee to screen all capital investment proposals above a certain dollar amount. Members of the committee are encouraged to attend professional meetings and expositions and are required to read and obtain advanced education in the states of the various arts affecting the company and its products. This committee is then both a source and screen for proposed capital investments and/or investment opportunities. The goal is an informed, organized, unified search for investment opportunities. The committee can also be a dispassionate viewer of all investment requests above a certain figure and thus ensure that all requests will have an equal ear.

Committees can offer problems. A camel, it has been said, is a horse designed by a committee. They can be bureaucratic, slow-moving, and prone to politics. But our solutions to the problems encountered in this step are limited. What impresses me is that more seminar participants than not testify to the effectiveness of such committees in their company as a reviewing, screening, and ruling group.

The Postaudit

The fourth and last step of capital investment is the follow-up step that seeks to determine (1) how much, in total, the investment actually cost the company and (2) if savings are involved and were projected, what actual savings have been attained.

The steps taken for the postaudit and the records that have to be collected depend on the given situation. But the follow-up action has to incorporate the following steps:

1. Complete actual costs of acquisition and installation are collected and compared with the original cost estimates.

2. Differences between estimates and actual costs are investigated. This is a valuable way to learn from your mistakes and to develop better capital investment estimating skills for future investment studies.
3. Actual operating performance and results are collected.
4. Actual operating performance and results are then compared with the original estimates of performance that were expected. If discrepancies exist, they are investigated and where possible, corrected.
5. The return on investment actually being achieved is calculated for comparison with the projected return upon which the decision to invest was originally made.

Since these five steps usually represent a lot of work, postaudits are done only for the larger capital investments.

Normally the postaudit should be done by a group other than the one that made the original estimates of investment costs and savings. An ideal agency to perform the postaudit is the controller's group, but one difficulty is that the follow-up studies on capital investment often require a certain degree of technical knowledge and shop operation know-how. The controller's staff sometimes do not know what questions to ask and what costs and savings to collect on this follow-up work. Certainly, the work demands more than simply an accounting-record approach if it is to be done completely and properly. The more progressive controller's staffs have such technical talent, like an industrial engineer or two in the group.

The harsh fact about this postaudit step is that only a very small percentage of our companies actually do it. I base that statement on my own experience and polls of those attending estimating seminars. Managers from large divisions of some very famous companies admit that it is not done in their enterprises. But how can we learn unless we identify our mistakes?

Bibliography

Clugston, Richard: *Estimating Manufacturing Costs,* Gower, London, 1971. British terminology.

Dumas, Lloyd J.: Parametric Costing and Institutionalized Inefficiency, *American Institute of Industrial Engineers Transactions,* June 1979. An outstanding and realistic article with good references. Well worth reading by every estimator involved in conceptual estimating.

Gallagher, Paul F.: *Parametric Estimating for Executives and Estimators,* Van Nostrand Reinhold, New York, 1982. Particularly pertinent for estimators in defense industries.

Ostwald, Phillip F.: *Cost Estimating for Engineering and Management,* Prentice-Hall, Englewood Cliffs, N.J., 1974. Very comprehensive, apparently aimed at college teaching. Bibliography and problems after every chapter but unfortunately no answers.

Tucker, Spencer A.: *Cost Estimating and Pricing with Machine Hour Rates,* Prentice-Hall, Englewood Cliffs, N.J., 1964.

Vernon, Ivan R. (ed.): *Realistic Cost Estimating for Manufacturing,* Society for Manufacturing Engineers, Dearborn, Mich., 1968. One of the best. Heavy concentration on metalwork. Should be in every metalworking estimator's library.

Wilson, Frank W. (ed.): *Manufacturing Planning and Estimating* Handbook, McGraw-Hill, New York, 1963.

Detailed Estimating Data

Hadden, Arthur A., and Victor K. Genger: *Handbook of Standard Time Data: Machine Shops,* Ronald, New York, 1954. Speeds and feeds can be updated by reference to the

Machining Data Handbook (see below). Still valuable to an estimator for elemental descriptions, allowances, format of approach, and other matters on estimating technique.

Hartmeyer, Fred C.: *Cost Estimating Data,* Ronald, New York, 1964. Only published time standards for electronics that I know of. May be usable if you have none of your own, but use with care and possibly with revisions downward.

Machining Data Handbook, 2d ed., Machinability Data Center, 3980 Rosslyn Drive, Cincinnati OH 45209. Most up-to-date, comprehensive metalworking speeds and feeds available.

Matisoff, Bernard S.: *Handbook of Electronic Manufacturing Engineering,* Van Nostrand Reinhold, New York, 1978. Manufacturing labor standards are given on pp. 57–98.

National Technical Information Services (NITS), U.S. Department of Commerce, 5285 Port Royal Road, Springfield VA 22161. Publish craft standards, including sheet metal and welding; ask for list of Engineering Performance Standards.

Tools and Dies

Harig, Herbert: *Estimating Stamping Dies,* Harig Educational Systems, 11802 N. Blackheath Road, Scottsdale AZ 85254.

Nelson, Leonard: How to Estimate Dies, Jugs and Fixtures from a Parts Print, *American Machinist/Metalworking Manufacture,* Sept. 1, 1961 (Special Report 510). An excellent work; participants in my seminars have found it to be eminently usable.

Learning Curve

Cochran, E. B.: *Planning Production Costs: Using the Improvement Curve,* Chandler, San Francisco, 1968. The most comprehensive learning-curve book I know of; unfortunately out of print. If you come across a copy, new or used, buy it.

Gallagher, Paul S.: *Project Estimating by Engineering Methods,* Hayden, New York, 1965.

Nanda, Ravinder, and George L. Adler (eds.): *Learning Curves: Theory and Application,* American Institute of Industrial Engineers, Work Measurement and Methods Division, Monograph Series 6, Atlanta, Ga., 1977. Collection of 18 articles on the learning curve. References at the end of each article total up to the best list of learning-curve references I know of. Some pretty heavy math at times. Although more and better exposition of real-life situations are called for, this remains a real service for students of the learning curve by the AIEE and the editors.

Estimating in Uncertainty

Burch, Jim D.: Cost Estimating with Uncertainty, *Industrial Engineering,* March 1975.

Deldecq, A., A. Van de Ver, and D. Gustafson: *Group Techniques for Program Planning,* Scott-Foresman, Glenview, Ill., 1975.

Van de Ver, Andrew: *Group Decision Making and Uncertainty,* Kent University Press, Kent, Ohio, 1975.

Capital Investment

Capital Investment, parts 1 and 2, two collections of 14 articles each from *Harvard Business Review;* available as nos. 21045 and 21046 from Reprint Department, Harvard Business Review, Boston MA 02163.

Lucey, T.: *Investment Appraisal*, Institute of Cost and Management Accountants, 63 Portland Place, London W1 H4 A8, England.

National Association of Accountants: *Financial Analysis to Guide Capital Expenditure Decisions*, Research Report 43, July 1976; *Return on Capital as a Guide to Managerial Decisions*, Research Report 35, 1959. These reports are basic, clear, and inexpensive; well worth having in your library.

Rate of Return: Toughest Measure of the Manager, *Factory Management and Maintenance*, November 1958.

Terborgh, G.: Business Investment Policy, Machinery and Allied Products Institute, Washington, 1958.

Appendix
Professional
Groups in
Estimating

American Association of Cost Engineers, 308 Monongahela Building, Morgantown WV 26505.

American Society of Professional Estimators, 17046 Weber Drive, Chandler, AZ 85224

National Estimating Society, 904 Bob Wallace Avenue, Suite 213, Huntsville, AL 35801

Society of Manufacturing Engineers, 1 SME Drive, P.O. Box 930, Dearborn MI 48128.

Index

About the Author

Lawrence M. Matthews is a Certified Management Consultant who has served over 150 clients in his 35 years of professional practice, first as member and part owner of a New York City consulting firm and more recently as a private management consultant. His expertise covers a broad range of industries, in both the manufacturing and service sectors. He has conducted hundreds of managements seminars in this country and abroad. Mr. Matthews is the author of *Practical Operating Budgeting* (McGraw-Hill) and is a senior member of the American Institute of Industrial Engineers and the National Association of Accountants.